THE EARTH CARE ANNUAL 1993

THE EARTH CARE ANNUAL 1993

Edited by Russell Wild

 National Wildlife Federation

 Rodale Press, Emmaus, Pennsylvania

Our Mission

We publish books that empower people's lives.

RODALE BOOKS

The *Earth Care Annual* team:

Editor: *Russell Wild*

Research Editor: *Melissa Gotthardt*

Production Editor: *Jane Sherman*

Reprint Permissions Coordinator: *Michele Toth*

Book Designer: *Lisa Palmer*

Cover Photo: *David Hall*

Copy Editor: *Amy Kowalski*

Office Staff: *Mary Lou Stephen, Julie Kehs, Roberta Mulliner*

Vice President and Editor in Chief, Rodale Books: *William Gottlieb*

Other contributors to the *Earth Care Annual* project: *Lynn Donches, Christine Dreisbach, Janet Glassman, Bill Howard, Ben McNitt, Stacey Overman, Bob Strohm, Debora Tkac, Mark Wexler, Lenore Wild*

If you have any questions or comments concerning this book, please write:

Rodale Press
Book Readers' Service
33 East Minor Street
Emmaus, PA 18098

ISBN 0–87596–136–3 hardcover
ISSN 1054–0067

Distributed in the book trade by St. Martin's Press

2 4 6 8 10 9 7 5 3 1 hardcover

Contents

FOREWORD

By Jay D. Hair
President and Chief Executive Officer
National Wildlife Federation

How effective can any individual voice be in a world grown so complex and large, so bureaucratically impacted, so callous to change?

There was a time in our past when the single voice of a lanky Illinois lawyer aroused a determination to keep the nation whole when he said that a house divided against itself cannot stand.

More recently a lone voice in the political wilderness chided his government's timidity toward the spread of totalitarian racism as a posture "decided only to be undecided, resolved to be irresolute, adamant for drift, solid for fluidity, all-powerful to be impotent." In time, Winston Churchill emerged from the political wilderness to lead the fight against fascism.

We are not challenged today by civil war, nor by the goose-stepping armies of a dictator bent on world domination and racial purity. But those two voices from the past still speak to the most vital concerns of our own time.

Today the issue is not that our house is divided but that it is we who are at war with our home. The conflict is global, one in which our assaults upon the environment imperil its ability to sustain life and industry. We are in disharmony with the planet. It is a strategic division that cannot be sustained. Yet as the realization grows that we must reconcile this division, we are too often confronted by leadership that is decided only to be undecided, resolute to be irresolute, adamant for drift.

Within only the past year, disturbing new evidence has appeared of the discord within our home that is this Earth. We have learned that the ozone shield that protects against harmful ultraviolet radiation is being damaged at a rate 200 to 300 percent greater than previously believed. The damage is almost certain to extend over areas of North America and Europe during springs and summers. Research by scientists who have probed mountain glaciers in remote hinterlands and tapped into climate records locked deep underground has pointed to the conclusion that the first ripples of accelerated global warming may already be under way. Unsettling reports have described a rapid fall in the populations of songbirds in many parts of the nation. Scores of communities have become embroiled in controversies over the siting of landfills and incinerators. Many of these wind up in minority neighborhoods or on Native American lands and are used for garbage or hazardous waste hauled in from hundreds or thousands of miles away. In rain forests spread through the tropics of Asia and South America, species that we haven't even had time to name have disappeared into extinction in the clear-cutting and flames that consumed their once-lush habitat.

Something else happened during the past year, a milestone in the struggle to heal our riven planet. First scores, then hundreds and finally thousands of people—most of them without claim to official titles or government authority—became involved in the process that lead to the Earth Summit in Rio de Janeiro. They wrote a Rio Declaration, a fledgling document now, but one they hope will mature into an Earth Charter with the same strength as an earlier fledgling document, the International Declaration of Human Rights.

These same people, during months of interchange, debate, cajoling and discovery, contributed to the formulation of an agenda to guide environmentally sustainable development into the twenty-first century. At the summit itself many of the nations of the world committed themselves to the first international agreements to confront climate change, deforestation and the threats to global biological diversity.

The summit should be judged not on the basis of the perfection of its results but as a symbol that our awareness has finally assumed a proportion commensurate with the dimensions of the task before us.

That act, the individual recognition of the reality with which we must deal, is the essential prerequisite to restoring wholeness to the global environment.

That recognition occurs one person at a time. It can take myriad forms. It can be as simple as the choice between a product presented with less packaging rather than the one with more. It can be as complex as the years of painstaking planning that went into the reintroduction into the wild of a colony of black-footed ferrets in the prairies of Wyoming over the past 12 months. It can be as public as a petition drive or as private as a talk along the edge of a pond with your child.

It is above all a personal decision to do what one can do within the scope of one's influence.

That's what this book is about. People who have done what they could do within the scope of their own influence.

The National Wildlife Federation is pleased once again to be working with Rodale Press to bring this volume, the fourth in an annual series, to you. It is a collaboration, drawing from some of the finest environmental journalism produced over the past year, intended to contribute to making the 1990s the Decade of the Environment.

Two main features emerge from the stories you are about to read. These are accounts of women and men who found out how much they could do once they decided to do something. Their ultimate reach exceeded their initial grasp. These stories are also accounts of the amplifying and reverberating effect that is part of the synergy of becoming involved. Commitment begets commitment.

For the person who wants to join the journey to reclaim our environment, these accounts may serve as a primer. For those who have already embarked on that trail, they may serve as affirmation. In either case, they provide vivid examples of the potential that exists within each individual to make a positive difference.

In these pages are the accounts from our own time of those who have raised their voices and committed their action to reestablishing integrity to our global home. Uniformly, they depict individuals who drew their resolution from within and from their association with others similarly committed. These are people who realize that the matters before us are far too important to leave solely in the hands of official authority. They understand that if we are to convert drift into decision, the impetus for change must come from the people.

EARTH CARE
FOR A GREEN DECADE

CORAL REEFS

Soft and hard coral combine to form a reef off the Caribbean island of Little Cayman. Coral reefs are the tropical rain forests of the seas, teeming with color and life. (Robert Samuel)

THE GHOSTS OF CORAL PAST

By Betsy Carpenter with Charlene Crabb

Lucy Bunkley-Williams deftly pilots her motorboat through the glittering, azure seas off Puerto Rico's southwest coast. Terns soar overhead, occasionally diving for herring, while pelicans paddle nearby. But the marine scientist is focused on an underwater reef, which she surveys for warning signs of a mysterious plague that has struck twice in the last four years. Both times, the reef's richly colored corals turned a ghostly white. "It was frightening," she recalls. "We couldn't imagine what had caused such a dramatic and rapid change."

Bunkley-Williams, a researcher at the University of Puerto Rico, has good reason to fear for the health of the reef. In the past decade, widespread "bleaching" of coral has become all too common—especially in the Caribbean, where seas have been unusually warm—and fear is growing that reefs may be the first major ecosystem disrupted by global climate change. Last fall, corals sickened from the Flower Garden Banks off Texas to Curacao. Earlier this year, coral in the waters of French Polynesia blanched. Reefs off the southern islands of Japan are beginning to bleach, and now, fresh reports from the Florida Keys suggest that another wave of bleaching may be sweeping the Caribbean. "Corals may prove to be canaries in the mine," says Bunkley-Williams, "the first indicator of global warming."

> Corals may prove to be canaries in the mine; the first indicator of global warming.

Despite their size, coral reefs are in fact among the most sensitive of nature's domains, existing within a narrow range of temperature, nutrient and light conditions. Most of a reef's vast structure is limestone, laid down over millions of years by a living "skin" of tiny, tentacled coral animals, called polyps, that build the reef as they grow. Polyps are colorless, but they harbor in their tissues armies of green and brown

3

single-celled algae that lend the reef its vibrant hues. The algae also feed the polyps a feast of sugars made through the process of photosynthesis.

Biological Eviction

Bleaching occurs when coral expels its algae, leaving the limestone skeleton visible through the translucent polyps. For decades, divers have reported small, scattered outbreaks of bleaching, the consequence of battering by a variety of environmental insults. In the Philippines, for instance, reefs blanch when aquarium-fish collectors capture iridescent blue tangs and angelfish by squirting poisonous sodium cyanide into reef crannies where the fish hide. In the Caribbean and the Florida Keys, hordes of sport divers trample delicate corals. Oil spills, hurricanes, pesticide runoff and the destruction of mangrove swamps, which filter silt and excess nutrients from coastal waters, have also bleached corals. But the current outbreaks are unprecedented in their severity and geographic scope. "Our reefs are in grave peril and are disappearing at an alarming rate," says Robert Wicklund, director of the Caribbean Marine Research Center on Lee Stocking Island in the Bahamas.

Researchers first saw whole reefs devastated by warm seas in 1983, during what is known as El Niño, a periodic and poorly understood warming of the equatorial eastern Pacific Ocean. The 1983 episode was particularly severe, and coral from Costa Rica to Ecuador turned stark white. Fully half of the coral colonies off the west coast of Panama died, and a species of fire coral that lives only in the eastern Pacific became extinct.

Signs of Sickening

Recent studies of the Caribbean have supported the hypothesis that abnormally warm oceans provoked the recent outbreaks. Wicklund and his colleagues began monitoring water temperatures on Lee Stocking Island's mile-long reef after a 1987 outbreak in which four-fifths of the coral sickened and an immense thicket of staghorn coral died. The 600-acre island is remote and uninhabited, apart from the research station, suggesting that pollution could not have poisoned the reef. In 1988 and 1989, Wicklund found water temperatures at or below the normal summertime high of 30°C (86°F), and the reef didn't bleach. But in August 1990, the sea warmed to between 31 and 32°C (88 and 90°F) for 18 days, and just a week later corals turned white.

Unfortunately, few research stations have kept accurate, long-term temperature logs, so some scientists have turned to satellite data for clues about bleaching. Alan Strong, a physical oceanographer at the National Oceanic and Atmospheric Administration (NOAA), recently examined sea surface temperatures recorded by NOAA's weather satellites in the late summer and fall of last year [1990] and discovered that the water temperatures in virtually every area hit by bleaching had been unusually warm. Laboratory research, too, suggests that corals are extremely sensitive to even small temperature spikes in the summer when tropical seas already are as warm as soup. Marine biologists have induced bleaching in corals by bathing them in water that is just one to two degrees above the normal summertime high for several weeks, or three to four degrees for a couple of days. But scientists agree that it is still too soon to conclude that global warming or natural fluctuations in ocean currents are causing seas to heat up.

Trouble from Above

The thinning of the ozone layer, which shields living creatures from damaging ultraviolet radiation, may also bear some responsibility for the recent demise of reefs. University of Hawaii marine biologist Paul Jokiel set up miniature reefs in outdoor tanks and exposed the coral to differing amounts of UV light using special filters. He found that when coral was already stressed by warm water, even a tiny increase in UV radiation aggravated bleaching. "These stresses are additive," says Jokiel. "When a coral is living near the lethal limit of its temperature range, other stresses, even little ones, can easily push it over the top."

Why the fragile partnership between coral and algae dissolves when seas warm is still unknown. One theory holds that when warm oceans boost the algae's rate of photosynthesis, the polyp suddenly is overwhelmed by oxygen, a by-product of photosynthesis that is poisonous at high doses. Leonard Muscatine, a professor of biology at the University of California, Los Angeles, has observed that the "glue" binding polyp cells together weakens when warmed, and algae-containing cells are sloughed off. In any event, polyps shed their algae-containing cells in a sticky mucus that stains seas a yellowish brown.

Reefs are the marine equivalent of tropical rain forests, teeming with a profusion of living forms: undulating sea fans and sea whips, feathery crinoids, neon-hued fish and sponges, shrimp, lobster and starfish, as well as fearsome sharks and giant moray eels. All depend on the continuing industry of coral for habitat. Single, short bouts of bleaching typically do not destroy a reef's ecology. Polyps usually regain their algae and their color a month or two after temperatures subside, but while corals are bleached they stop reproducing, grow only slowly and are more susceptible to disease.

Reefs are the marine equivalent of tropical rain forests, teeming with a profusion of living forms.

But repeated bleachings can devastate both the reefs and the myriad creatures that depend on them, as is apparent in the Florida Keys, which have suffered through four outbreaks in the past decade. Though overfishing, changes in salinity and pollution have stressed the reef ecology, unusually warm seas have been a key factor in the demise of reefs, say Billy Causey, project manager for the Florida Keys National Marine Sanctuary. Many reefs in the Keys have been overrun by thick, fleshy mats of algae, which prevent new polyps from recolonizing the limestone skeleton. Dead stands of elkhorn, staghorn and brain coral are crumbling, destroying habitat for plants and animals. The ranks of angelfish, butterfly fish, tangs, grouper and snapper have dwindled, and in some places, brilliantly colored sponges and anemones have completely disappeared.

In the Keys, the bleaching season is again under way. During the last week of July [1991], several species of coral near Long Key, about 65 miles east of Key West, began to blanch as the water temperature climbed to 31°C (88°F). New reefs through- out the region look as if they have been dusted with snow. The damage to Florida's reefs is far from over.

From *U.S. News and World Report*, September 23, 1991. Copyright © September 23, 1991, *U.S. News and World Report*. Reprinted by permission.

 EARTH CARE ACTION

Loving the Reef to Death

By David J. Fishman

The fish must think we're crazy. Just 50 years ago, if they glanced through the surface of the sea, they never would have seen humans walking like ducks along the edge of boats, spitting gobs into silicone faceplates, chomping on rubber mouthpieces and hissing compressed air. Even 20 years ago, when the mass marketing of scuba training and its associated Day-Glo accoutrements began in earnest, these same fish rarely encountered more than a handful of these aliens on a typical day.

Today however, the fish that live in the nooks and crannies of coral reefs from Florida to the Red Sea bear witness as legions of these lead-bearing invaders, cans of Cheez Whiz stuffed in their neoprene pockets, come crashing through their crystal ceiling and crowd into their backyards.

There are now more than four million certified scuba divers in the United States alone. At least ten million more people are expected to learn scuba worldwide during this decade. Diving, one of the world's fastest-growing sports, now fuels a multibillion-dollar business of dive travel, expert certifications, computerized gadgets and sophisticated camera gear. With new airline service and luxurious live-aboard yachts, traveling divers have easy access to once-isolated parts of the world. Millions of American divers can now visit Belize, Bonaire and the Cayman Islands as routinely as the Florida Keys. Once-isolated destinations are quickly becoming scuba factories, enticing cash-rich divers to their previously quiet shores.

But with sport diving growing exponentially, marine biologists as well as an increasing number of people within the dive community are growing uneasy. The world's coral reefs are already besieged by a nasty concoction of ills:

unexplained outbreaks of disease and bleaching, sewage and agricultural runoff, oil spills, boat groundings, careless anchoring, overfishing and overcollecting and even dynamiting. Now coral reef ecologists are trying to understand the added effect of the growing hordes of underwater sightseers. On reefs all over the world, divers kick, grab and break corals, cover them with choking sediments and bang them silly by dragging high-tech consoles that hang below them as they swim. In addition, divers' seemingly harmless and good-natured interactions with reef critters may instead be deadly.

*O*nce-isolated destinations are quickly becoming scuba factories.

"When people go tromping through the national forest," says Alina Szmant, a coral biologist at the University of Miami's Rosenstiel School, "you get trails and pieces of litter, broken branches, people urinating and other damage. Divers in the ocean, however, are not limited to trails and they can do a lot more damage."

Sad Sights

Anyone who has spent time on a coral reef has seen the type of damage snorkelers and scuba divers can do, especially inexperienced ones. The sad sights include patches of dead, algae-covered coral on otherwise thriving heads, broken tube and barrel sponges and freshly cracked-off branches of delicate corals.

Improper buoyancy control, the hallmark of a beginning scuba diver and the chief cause of the physical damage done by divers, is almost always associated with a diver wearing too much lead weight. After they enter the water, overweighted divers smash down on the reef bottom. Without proper buoyancy in the water, divers knock into the corals on the reef, pushing off to raise themselves up or holding on to stay down. "Until they are taught how to master buoyancy control, these divers are menaces," says Dee Scarr, a marine naturalist and dive group leader on Bonaire. "I only take a few people with me at a time so I can watch how they're doing."

Unfortunately, most divers aren't escorted around the reef by a concerned naturalist. Rather, fleets of boats each stuffed with 20 or more divers shuttle to pristine reef sites where novice and expert alike wander the reefs in buddy pairs. In areas where reefs reach into shallow waters, scuba divers are joined by additional boatloads of snorkelers, many of whom are jumping into reef waters without any prior experience.

While it never ceases to amaze boat captains and divemasters, a lot of the people they take to the reefs have no idea that the corals beneath them are alive. Many of these divers surface with their legs and hands full of mucus and scraped-off coral tissue. "It all boils down to education," says Scarr. "People should have to take a course in local ecology when they come to a new destination, especially if they've never been on a reef before."

Stirring Up Sediments

To assess what type of reef visitor does the most damage to the coral, a marine biologist from the University of South Florida organized a summer-long study at one of Florida's most popular dive sites: the Looe Key National Marine Sanctuary. During the summer of 1989,

Helen Talge anonymously observed 206 divers and snorkelers as they swam underwater. Working in cooperation with sanctuary officials and the commercial dive-boat operators, Talge tallied the number of times divers bumped into, scraped or pushed off on the hard and soft corals. She also monitored how much sediment each diver stirred up.

On reefs all over the world, divers kick, grab and break corals.

Talge found that the profile of a diver who had the most damaging effect was a young male scuba diver wearing gloves. She also found that divers who racked up the most excessive number of negative interactions (the highest averaged one per minute over the course of a 30-minute dive) were in the water with a specific objective in mind: lobstering, shell collecting or photography.

Women, it seems are more inclined to treat the reef gently. Talge found that the profile of the most careful visitor was a woman snorkeler who dove without gloves. She also found that women scuba divers tend to stay in the sand channels between coral heads and observe the reef from a greater distance than men.

Considering that the average diver had more than half a dozen negative interactions with the coral and that more than one million people visit Florida's marine sanctuaries each year, even the slightest, most casual touch can have a devastating cumulative effect. The polyps of a typical hard coral colony are contracted during the day when the most divers encounter them, but even then there is a thin film of living tissue draping the stony skeleton. "Whenever you press on the corals with your hands or your feet you can very easily puncture these thin layers of tissue against the animal's hard skeleton," says Szmant. "This then exposes the corals—like someone cutting you down to your bone—and things like algae, bacteria and fungi can get in there and start an infection. Normally the corals will be able to recover from the isolated bit of damage, as they do from the occasional bites of grazing parrotfish, but if you have thousands of people wearing gloves doing this all over the place all the time, the corals might not be able to repair themselves."

The size of the coral wound can also be critical. According to James Porter, a zoologist at the University of Georgia, if the abraded area is larger than a few square inches, then there is only a 10 to 20 percent chance that the coral will be able to reoccupy the area before algae or another infector gains a foothold and prevents regrowth. And if the water conditions are right, the invader could spread to the rest of the colony.

The snorkelers Talge followed in her visitor study, while having seven times fewer interactions with the coral, created a different sort of problem. "Inexperienced snorkelers tread water in the shallows, stirring up huge clouds of sediment which then settle on the corals. And when they get tired they simply stand up to rest on a living coral head."

While hard corals can slough off the choking sediments stirred by a single snorkeler, they have more difficulty removing the sand if it covers the colony repeatedly. "Corals have evolved to deal with brief periods of sedimentation caused by storms, or a flurry of parrotfish feces," says Szmant. "But when sedimentation is constant, they wear themselves out producing the mucus to remove the sediments. And where corals are already under stress for other reasons, this could just be too much for them." Unfortunately for Florida Keys' corals, boats running out of John Pennekamp Coral Reef State Park bring hundreds of snorkelers to shallow reefs. Snorkelers also visit parts of the northern Great Barrier Reef, Eden Rock off Grand Cayman and the underwater trails of the U.S. Virgin Islands in similar numbers.

Preventive Action

In order to tip the scales back in favor of reef conservation, managers of protected reef areas and an increasing number within the dive industry are taking action to prevent further damage. "We need to concentrate on every source of impact regardless of its nature or how small it may seem," says Billy Causey, manager of the Looe Key National Marine Sanctuary. "We have a great deal of damage from snorkelers who stand up in the shallower areas behind the reef crest," he says. "And with the number of divers visiting the reef still growing, we realize that something has to be done."

Perhaps the most creative of the measures that Causey and Mike White, the manager of the larger Key Largo National Marine Sanctuary, have taken is the establishment of an underwater police force. On several days during peak visiting season, sanctuary officers don scuba gear and patrol the most sensitive parts of the reef. If they spot a particularly egregious diver they will approach him and, with the aid of an underwater slate, direct him to surface. On the surface the offender can be issued a $25 civil penalty called an Enforcement Action Report. More often, though, the rangers use the opportunity to educate the diver about the consequences of his or her actions.

"The vast majority of our sanctuary violators are doing damage unintentionally," says White. "That's why our law enforcement approach is largely educational. Every day our sanctuary officers are out on the water going from boat to boat handling out brochures and explaining the importance of our conservation efforts."

Diving Etiquette

Both Causey and White agree that although the reef patrol program creates a lot of publicity, the strongest part of their program is the support they receive from working closely with the area's commercial dive shops. It is through the dive shops and their dive boats that the majority of the visitors to the sanctuaries get to the reef. "We have to give them credit for initiating things like not allowing divers on their boats to wear gloves," says Causey. "They'll even rat on their passengers sometimes. A captain will call our officers over and say 'after the first dive this guy had coral mucus all over him—watch him on the next dive.' "

Also, corals in the Florida Keys are only protected within the limited stretches inside the two national sanctuaries and Pennekamp State Park. On some islands in the Caribbean, the managers have closed some of the more heavily visited dive sites for long-term recovery. Unless there is a significant expansion of the marine sanctuary program in Florida, however, there is simply not enough room to rotate the available dive sites.

Lauri MacLaughlin, the education coordinator for Looe Key, believes that the scuba-certifying agencies need to be encouraged to put more reef etiquette into their certification programs. "Teachers need to spend more time than the five-day slam courses they offer down here in order to attain the necessary skills," she says. "It takes more time than that simply to master buoyancy."

Reef Awareness

In response to this growing concern, the Professional Association of Diving Instructors (PADI) is attempting to increase the reef awareness of its divers. Last January [1991] it introduced Project AWARE, an acronym for Aquatic World Awareness, Responsibility and Education. The goal of PADI's ten-year program is to instill in its students a conservation ethic while also redoubling its efforts to train divers with sound skills. "We want our divers to be welcome guests wherever they venture," says Barry Shuster, a PADI spokesperson. For its part, the National Association of Underwater Instructors (NAUI) is encouraging members to get a "buoyancy check" as part of a regular refresher program. Gary Caldwell, NAUI's program coordinator, believes northern training can make divers dangerous to the reefs. "It's hard to do damage to the granite lining the dive sites where many of these divers train," he says.

Of all the divers who visit the coral reef, the underwater photographer is most often cited by experienced observers for causing unintentional coral damage. "In their zeal to get the best shot, underwater photographers can create problems," says Stephen Frink, one of the world's most prolific underwater photographers. "I know that when I started diving I was perhaps more careless. Part of it was ignorance: I thought that the ecosystem was inexhaustible. Part of it was the vanity of thinking a photo might be so significant it will communicate the fragility of the reef and help it survive."

Renowned underwater photographer Christopher Newbert is also a dive-tour leader known for his strict conservation policy. "I make a speech at the beginning of a trip on how divers should position themselves for a picture," say Newbert. "And we tell them outright before we even go out that if we see them damaging the reef in any way, we are going to restrict their diving activities."

Newbert feels that an even more neglected problem related to photography is the handling of wildlife. "The media, especially the dive magazines, needs to be attacked for promoting manhandling of marine life," says Newbert. "Every cover shot is a picture of some diver strangling a pufferfish—stressing it until it puffs up, handling an octopus, sticking a shrimp on his mask or petting an eel.

"These pictures spawn a lot of imitators and the neophyte underwater photographer figures, 'Oh, this is the state of the art, to have a girl in a bikini with matching eye makeup and lipstick riding a turtle,' " says Newbert. "With very few exceptions, you can't do things like ride manta rays or pet stingrays without frightening and overly stressing the animals."

It's obvious from the frustration of those who know the reef best that divers and snorkelers need enlightening. The coral cuts permanently etched into our fins can no longer be ignored. We must learn to be gentle divers, to limit the evidence of our presence and passage.

The fish may still think we're crazy, but at least we'll know we're doing all we can do protect their way of life.

From *Sea Frontiers*, March 1991. Reprinted by permission.

Stupid Diver Tricks

Scuba divers and snorkelers can do a lot of unintentional damage on a coral reef. By learning what *not* to do, you can make your next visit to a coral reef enjoyable for yourself and safe to the environment. Here's a list of what we call "stupid diver tricks":

➤ Improper buoyancy control: too little weight and divers will grab onto the fragile corals to stay down; too much weight and they drop on top of the reef.

➤ Divers churn up a lot of sediment, which can choke corals and prevent sunlight from reaching the colony.

➤ Divers trail equipment—hoses, gauges, heavy consoles, extra regulators—below them that drag on the corals.

➤ Divers forget they are wearing tanks on their backs and repeatedly scrape against the coral and sponges, trying to fit through spaces that are too small.

➤ Divers kick the coral with their long fins. When in a vertical position, divers tend to bicycle kick wildly. Instead they should turn horizontal for a gentler, more graceful posture.

➤ Photographers stabilize themselves by grabbing. They can also be blind to the reef, their eyes glued to a viewfinder.

➤ Instructors bring beginners to pristine reefs for training.

➤ New divers thrash around when they panic, hitting corals and stirring up sediment.

➤ Divers often jerk themselves free when snagged, instead of gently freeing themselves from what caught them.

➤ Many divers are trained in northern climes, where the granite lining the ecosystem is much more forgiving.

➤ Many divers wear gloves and Lycra body covering, which encourage contact with the reef and make mistakes easier.

➤ Divers often overturn rubble looking for large sea life, leaving small residents exposed and vulnerable.

➤ Divers handle or even sit in large sponges, breaking the thin, growing edge of the sponge and causing permanent damage.

➤ While feeding fish, divers often leave behind plastic bags or cans of Cheez Whiz, when indeed they shouldn't be feeding the fish at all.

➤ Divers collect dead shells, forgetting that these are potential homes for many different animals.

➤ Divers can prevent young corals from gaining a foothold by disturbing the apparently barren bottom.

➤ Diver's bubbles can scour vertical surfaces (walls), ledges and overhangs. Bubbles can sweep away small critters and fracture some fragile corals.

➤ During night dives, divers wake sleeping fish with their bright lights, blinding them and causing them to rush into obstacles and injure themselves.

WHAT YOU CAN DO

An organization called The Ecotourism Society provides information on environmentally responsible travel. If they can provide you with information on ecotourism, or if you would like to offer feedback on your traveling experiences, contact them at 801 Devon Place, Alexandria, Virginia 22314; (703) 549-8979.

➤ Night divers' limited vision and greater anxiety make them likely to bump corals whose polyps are extended for feeding.

➤ Divers ride sea turtles, forgetting that the animal might have been on the way to the surface to breathe.

➤ Divers hold seahorses, frogfish and other docile wildlife, not realizing that too much contact and handling—especially when passing the animal around—can stress and possibly kill the animal.

➤ Sea urchins are killed to feed fish, but these grazers help control algae populations and keep corals healthy.

➤ Divers chase pufferfish, forcing them to inflate, not realizing that this is a sign of stressing the animal. Handling the fish can remove their essential mucus.

➤ Too many divers insist on visiting "celebrity sites," many of which now need time to recover.

From *Sea Frontiers*, March 1991. Reprinted by permission.

Reef Raiders

By Frederic Golden

When I was a youngster, the only place where you were likely to encounter a tank full of tropical fishes, apart from the local aquarium, was in your dentist's waiting room or in the dimly lit cocktail lounge of some fancy restaurant.

Of course, if you were determined to keep such creatures as pets, you could do as I did: buy goldfish at the local Woolworth's. These denizens of fresh waters weren't very exotic or tropical, but they were the next best thing. No fancy pumps or gadgetry were needed; only a simple bowl, perhaps with a garishly colored stone castle, and an occasional cleaning and change of water would keep Clio and Henry reasonably perky.

Nowadays, fish collecting is much easier, so a rich variety of colorful creatures—not just goldfish but tropical marine species—is available. What's more, the technology and special know-how to keep these saltwater animals alive at home are available to the amateur: large silicone-sealed tanks as leakproof as a Navy sub; filtration and bubbling systems that would be the pride of any distillery; bacteria-loaded gravel that can detoxify even the most acrid fish-killing wastes.

Thanks to such high-tech innovations, plus an increased interest in the sea stirred by television nature shows and the proliferation of public aquariums around the country, many Americans have become aquarists, as these hobbyists are called. In numbers, at least, fishes are the nation's favorite pets. Industry officials say that there are now 200 million fishes in American homes, outnumbering both dogs and cats. And if you ask any of their owners, they'll tell you they're as much of a joy to own.

In my ichthyological innocence, I had to be content with goldfish. Today's marine aquarists have their pick of the reef. Or so it seems. Some 200 of the 4,000 small marine fish species that dwell on coral reefs are found in pet shops. And the more that's spent on it, the rarer the specimen is likely to be. I paid only a few dollars for my fish and tank; now aquarists often shell out $1,000 or more to get started in their hobby.

Little Fishes Equal Big Business

On a national scale, such purchases translate into a big business. Americans spend an estimated $1 billion a year on aquarium fishes and the assorted paraphernalia and supplies needed to take care of them. This has spawned a thriving industry, with a large cast of characters. They include the freshwater fish farmers and collectors; marine fish collectors who scour coral reefs for specimens; the wholesalers, distributors and shopkeepers who market them;

Prize held aloft, fish collectors in the Philippines prepare to head back to shore. Specimens collected using cyanide often don't survive the trip to market. (Tom Stack and Associates © Jeff Foott)

and the sundry manufacturers and suppliers who provide the food, drugs and equipment to keep the fishes alive.

> *F*ishes are the nation's favorite pets. Industry officials say that there are now 200 million fishes in American homes, outnumbering both dogs and cats.

In addition, the hobby has created a demand for many support services. You'll find videos as well as books and lectures on fish care. Half a dozen specialty magazines exist solely to cater to needs of aquarists. Aquarium fish clubs and societies are flourishing in many places. Members exchange information, equipment and even fishes ("I'll give you three of my wrasses for that tang of yours").

As hobbies go, home aquariums are not only great fun but also highly satisfying, especially in these environmentally conscious times. For the marine collector, they encourage both an appreciation of, and intimacy with, some of the more spectacularly dramatic life forms in the sea. And yet, for all that can be said in behalf of this attractive hobby, it is clouded by troubling questions. Has the love affair with tropical marine fishes become so strong that we're overwhelming the object of our affections? By taking so many fishes out of the sea in order to admire them close up, are we threatening their survival?

One would think there is very little reason to worry. Fishes, after all, are among nature's most prolific creatures.

"They're extraordinarily fecund," says John McCosker, director of San Francisco's Steinhart Aquarium. "They lay eggs by the millions. So long as they're properly captured, they're a renewable resource." Unfortunately, he adds, all too often they are not taken out of the sea in an environmentally sound manner. In many parts of the world, collectors are destroying the very habitat on which the fishes depend.

If marine display fishes could be bred on fish farms, there would be no problem. Yet except for a few compliant creatures such as damselfish, which are being raised on farms in Florida and Georgia, the overwhelming majority of tropical marine fishes come from the sea. They are collected from coral reefs in the Caribbean, the Bahamas, Mexico, Sri Lanka, Australia and the tropical Pacific. Florida and Hawaii, our only states with tropical reefs, forbid unlicensed fishing altogether. But in most places, collecting is either unregulated or not regulated effectively. Especially in Third World countries, the source of most species, there is always a risk of overexploitation, not just from collection of the fishes themselves but from the destruction of their critical reef habitat.

Philippine Treasure

Nowhere is the danger greater than in the Philippines. Its thousands of miles of coral reefs are one of the world's natural treasures,

with the highest density of marine species in the world (2,177 according to one recent count). Peter J. Rubec, president of the International Marinelife Alliance (IMA) USA, an environmental organization that has campaigned aggressively for protection of the Philippine and other reefs, says: "No other country can match the diversity of colorful species desired by the marine aquarist."

Because of the Philippines' natural abundance, as well as its low labor costs and accessibility by major airlines, it has virtually cornered the tropical fish market. Rubec estimates that as much as 80 percent of the marine fishes sold worldwide originate in the Philippines. This trade—mainly with the United States—is an important revenue source for the politically and economically troubled nation, producing earnings of $10 million to $20 million a year.

But the exports are taking a heavy natural toll. As anyone who has tried it knows, collecting small reef fishes is something of an art. At the first sign of threat, the skittish creatures scoot for cover in the reef's many nooks and crannies. "You have to be smarter than they are, you have to anticipate their moves," says Bruce A. Carlson, director of Hawaii's Waikiki Aquarium, who has tried his collecting skills in the lagoons of Fiji. "It's a tough business. You may find yourself working seven days a week to make it pay." Many fish collectors never do.

Knock-Out Cocktail

In the early 1960s, Filipino fish collectors learned that they could greatly improve the odds with chemical help. They began squirting their elusive prey with a dose of sodium cyanide. The poison killed some of the fishes instantly. But if the collector tempered the dosage, most of his targets were merely stunned. In their paralyzed state, they could easily be gathered up and carted off in a boat. Eventually, the fishes would revive and show no obvious ill effects from the whiff of cyanide.

But their apparent good health could be deceptive. Typically, the poisonous cocktail consists of sodium cyanide powder or tablets dissolved in water. (A few collectors used less-expensive chlorine.) Rapidly absorbed through the gills as hydrocyanic acid, the cyanide solution is converted by the fish into a compound called thiocyanate. Once scientists thought thiocyanate was largely harmless to fishes. But experiments with thiocyanate-laced trout have shown it makes them dangerously susceptible to stress. Just a bump or a slight exertion can trigger convulsions. The fish gasps wildly and rapidly loses its equilibrium and buoyancy. These episodes are invariably fatal.

Chemical Killers

Dealers and aquarists have long observed such cases of sudden death syndrome in their tanks. "You think you have a healthy fish," says ichthyologist Don E. McAllister of the Canadian Museum of Nature in Ottawa. "But then it dies for no obvious reason. And you ask yourself, 'What did I do wrong? Did I feed it enough?'"

Now the answer seems clear. Though the precise chemistry of such deaths remains under study, the link with cyanide-collecting is no longer in doubt. Subjected to stress, the cyanide-dosed fish converts the thiocyanate in its system back into hydrocyanic acid, which then acts on the neurological centers and sets off the convulsions. The cyanide also interferes with the oxygen-carrying capacity of hemoglobin in the fish's blood and disturbs the liver's vital enzyme activity. "Even if it looks all right, a fish that has been dosed with cyanide is a very unhealthy creature," says McAllister. "Almost any stress at all can kill it."

There are other, less stressful chemicals that can be used for collecting aquarium fish. Dilute quinaldine and ethyl alcohol, the only chemicals that can be legally used in the state of Florida, have not been associated with sudden death syndrome or any other deleterious side effects. Quinaldine is considered an anesthetic rather than a toxin like cyanide. When properly diluted and administered, fish can be kept sedated for extended periods. Most collectors who use quinaldine report 100 percent survival of their fish.

Stress is, of course, a fact of life for a display fish from the moment a collector eyes it on the reef, whips out his kitchen-type squeeze bottle and unleashes a cloud of the poison. Even if the creature survives the blast, it is likely to be stressed further as it is hauled to the surface (usually without decompressing its swim bladder), tossed into a holding tank and trucked to Manila. There it will be jammed into a tank with other specimens waiting to be air freighted to the United States. Still more transfers and jostling face the creature upon arrival in the United States as it makes its way from importer to pet shop and finally into a hobbyist's tank.

Even careful handling produces a high number of casualties during this globetrotting. Studies show that as many as three-quarters of the targeted fishes die immediately at the point of collection (not to mention untold numbers of larvae inadvertently caught in the chemical cloud). Of the survivors, up to half perish before they ever reach Manila, the Philippines' fish export center. There shippers lose about a third of their fishes; handlers in the United States experience about the same casualty rate. For every fish that makes it safely from reef to aquarist's tank, 20 or more have been sacrificed to get it there. "One is hard-pressed to think of a more biologically expensive hobby," says Massachusetts veterinarian Vaughn Pratt, IMA's cofounder.

Profitable, But Illegal

In response to the pressure from abroad, mostly from IMA's American and Canadian members, the Philippine government decreed cyaniding illegal several years ago, but the practice continues because it seems to yield such high profits.

Under a federal law known as the Lacey Act, the U.S. Fish and Wildlife Service can seize animals that have been illegally caught—in violation of international agreements on endangered species, for example—but it has never felt it could stop imports of cyanide-dosed fishes since they have always arrived properly certified by the Philippine government, which doesn't have an adequate test to identify them. IMA is trying to provide one.

To ensure the health of their fishes, public aquariums usually contract directly with their own collectors or agents. "We get many of our tropical marine fish from the Caribbean from people we know and trust," says San Francisco's McCosker. "When they do their collecting, they're going to do it with hand nets, not cyanide. They're not going to damage the reef or hurt the local population."

Ironically, the ideas for using cyanide to stun fishes originated with American fisheries researchers in Illinois who discovered in the 1950s that the chemical could be useful in cleaning Great Lakes' fishes. A collector in the Philippines heard about the technique. And when his catches began soaring, word of the cyanide-assisted fishing spread like a typhoon across the archipelago. Novice divers with only scant fish-catching experience on the reefs suddenly became large producers, thanks to cyanide. In Manila, the number of tropical fish exporters soared from only three in the early 1960s to more than three dozen a decade later.

While cyanide-induced trade brought immediate economic benefits for some people, it set off an ecological time bomb.

But while this cyanide-induced trade brought immediate economic benefits for some people, it set off an ecological time bomb. Though the collectors aim the cyanide at the fishes, some of it inevitably settles on the reefs. Biologically fragile, coral heads represent a precarious symbiosis between the individual animals in the colony (polyps) and the tiny algal cells that they feed on. To remain healthy, the coral requires the right mixture of sunlight, temperature and nutrients. Any imbalance not only damages the reef but also unravels the intricate food web that sustains the reef's rich life, including the tiny fishes so coveted by marine aquarists.

McAllister, head of IMA's very active

Canadian wing, argues that destroying the ocean's coral reefs is as biologically disastrous as the devastation of the rain forests. "A coral reef is one of those magical places, a hot spot for biological diversity and density, where you'll find fish of all sizes, as well as crabs, shrimps and lobsters, all of them living in marvelous balance." Too few people and nations appreciate the full significance of the reefs to ocean life, he says.

In the Philippines, coral-reef fishes provide half the protein in the local diet and are the direct source of employment of some 600,000 people involved in small-scale fisheries. A healthy reef can produce more than 90 tons of fish a year per square mile, among them many aquarium species. But once the reef starts to degrade—from cyanide or other damage—production quickly slumps, often to as little as a seventh of the original catch.

Temporary Damage Only?

Defenders of cyanide-fishing say that spraying by only a thousand or so fishermen—the number of collectors in the Philippines—can hardly have a major ecological impact on the archipelago's thousands of miles of reefs. They argue that, at worst, the cyanide causes only temporary local patches of discoloration—a result of the chemical's bleaching effect—from which a reef rapidly recovers. But McAllister strongly disagrees. "They pretend it's only an aesthetic issue," he says, but studies show that repeated doses leave reefs so badly damaged they eventually die off.

Steve Robinson, a professional fish collector from Los Angeles, has been retraining Filipino fishermen to use hand nets rather

than cyanide. He has done simple calculations to show them what havoc cyanide causes. A single collector, Robinson points out, squirts about 50 coral heads a day. Thus, 1,000 collectors working an average of 225 days a year would dose more than 12 million coral heads annually. Actually, says Robinson, the cyaniding is probably even more widespread, since it's also used to catch food fish, both for domestic consumption and for export to Hong Kong, where fish bought while they are still alive are believed to confer long life.

A healthy reef can produce more than 90 tons of fish a year per square mile, among them many aquarium species. But once the reef starts to degrade—from cyanide or other damage—production quickly slumps.

Cyanide is also highly dangerous to its users. It has killed a number of fishermen who have accidentally swallowed the toxic chemical or breathed the gas emitted when cyanide contacts moisture in a bag, and young men with cyanide burns on their bodies are not an uncommon sight in fishing villages. Some villagers have also died from eating fish that still had heavy traces of cyanide in their stomachs. "By now a million kilograms of cyanide have probably been sprayed on coral reefs, which, in theory, is enough to kill just about everyone in the Philippines," notes McAllister grimly.

Although cyanide fishing is illegal in the Philippines, it yields such high profits that campaigning against it is risky. Local officials who have attempted to enforce the ban have been threatened with death, as have environmentalists such as Robinson. "Yeah, I've had my share of warnings," he acknowledges, "but I won't let them stop me." Robinson, 36, who admits to a "love affair" with the Philippines, has been visiting its rural fishing villages since the early 1980s, trying to stop the use of cyanide.

Yet cyaniding isn't the only menace to the reefs. An even more popular fish-catching technique among Filipino as well as other Third World fishermen is dynamiting—stunning fishes with explosives. Almost any blasting material will do. In Indonesia, for example, "fish bombs" are made from the cordite in World War II shells that occasionally wash up on the beaches. The concussion wave not only stuns or kills fishes, it destroys large patches of coral.

Although cyanide fishing is illegal in the Philippines, it yields such high profits that campaigning against it is risky. Local officials who have attempted to enforce the ban have been threatened with death, as have environmentalists.

Perhaps the most devastating catching method, introduced into the Philippines by the Japanese, is *muro-ami* (a Japanese term that means roughly "thumping fish"). It's a good description of the technique: A veritable army of swimmers, from 50 to 300 strong, enters the water, each holding a weight the size of a man's head. Jerking the lines up and down, they pound on the reefs to scare the fish out of their niches and into a large waiting net. The practice not only destroys the reefs (from the impact of the weights) but also exploits the fishers, many of them youngsters, often only nine or ten years old, who live for many months at a stretch aboard crowded, ill-ventilated barges. In 1986 the Philippines banned muro-ami fishing, but the decree has never really been enforced.

Reefs under Attack

Apart from illegal fishing practices, reefs are also threatened from the land, especially by the clearing of tropical rain forests. As the trees are replaced by farms, mines or other human pursuits, the runoff increases, clouding the water with pollutants and sediment. Among other things, this reduces the amount of precious sunlight that falls on nearby coral, imperiling its survival. A few years ago, a study indicated that three-quarters of the coral reefs in the Philippines were in only poor or fair condition. "The situation has probably gotten worse since then," says McAllister, who has been trying to wake up governments to the devastation.

Elsewhere, he says, coral reefs are faring no better. On the Caribbean coast of Costa Rica, some 80 percent of the reef-building corals are believed to be dead as a result of such mistreatment. In Tanzania, East Africa, the reefs have been extensively damaged by fishing with dynamite.

IMA has long argued that keeping coral reefs healthy pays off. But, according to current estimates, collectors continue to pour about 150 tons of cyanide a year on Philippine reefs, even though they are destroying breeding and feeding grounds of their fish. "The practice makes about as much sense as a dairy farmer in

Wisconsin killing off his hay fields with herbicide and burning down his barn," says McAllister.

Last year [1990], with an initial grant from the Canadian government, IMA began its Netsman Project, a Peace Corps–type program aimed at teaching Filipino fishermen how to collect reef fishes without cyanide. It's been an uphill struggle, however. Most of the collectors are very poor, earning only a few hundred dollars a year, and are not always eager to jeopardize their catch with strange techniques. The program was initially opposed by the big exporters in Manila, and the American pet industry was largely indifferent (though both have now come around in support of it). "It was a typical case of denial syndrome," McAllister says. "They refused to recognize the threat to their own business."

Robinson, the program's field director, knew from his own experiences in the Philippines the resistance he would encounter. But as "an old peacenik" (his words) with a special interest in the Third World, he has developed a special rapport with rural Filipinos.

He understands their language and how to appeal to their self-interest. His argument is simple: Hand-collecting saves you the cost of cyanide ($6 or $7 a bottle), reduces your fish mortality rate to almost zero and just about doubles your catch. He also emphasizes that by protecting the reef you're ensuring good harvests in the future.

Teaching
a Better Way

An experienced scuba diver, Robinson goes with his students into the water, shows them how to prod the reef gently with poles or sticks and sweep up their catch with homemade nets. He teaches them how to decompress the fishes (by carefully puncturing their swim bladders with a needle), so they won't bloat when they're brought to the surface. He also demonstrates proper packing to get the fishes to Manila in good health, thereby ensuring the fisherman his fee. (One tip: Don't feed the fishes beforehand so they won't be poisoned by their own wastes.)

During the week-long courses, Robinson is assisted by specially trained Filipino tutors who he hopes will later spread the message on their own. In the first round of instruction under the Netsman program, Robinson taught 125 cyanide fishermen in towns of northern Luzon. They were remarkably receptive. Whenever they netted a fish, which was often, they would surface and jubilantly shout "wala sodium, wala sodium!" (no sodium, no sodium). At the graduation ceremonies, the fishermen proudly surrendered their cyanide bottles for certificates attesting to their new status as noncyanide collectors.

*A*t the graduation ceremonies, fishermen proudly surrendered their cyanide bottles for certificates attesting to their new status as noncyanide collectors.

"It's ironic that an American is teaching Filipinos how to fish," says Robinson, adding with characteristic hyperbole: "We're the worst fishermen on the planet. We depend on gear, not talent, while they depend on their tremendous skill and no gear. (Most Filipino collectors can't afford any breathing apparatus.) When I'm teaching them they learn so quickly, I get worried about how quickly they're going to make me obsolete."

That's not likely to happen very soon, however. Despite the recent hints of progress in the Philippines, reef populations still remain in serious danger from cyaniding and other activities. In another protective step, IMA soon hopes to introduce a simple, inexpensive bioas-

say by which Filipino inspectors will be able to test quickly whether fish have been exposed to cyanide. But the broader issue remains. "That's continued education at both ends of the pipeline," says IMA's Pratt, "among those who catch the fish and those who export them to us."

From *Sea Frontiers*, February 1991. Reprinted by permission.

ENDANGERED SPECIES

A green turtle swims lazily off the coast of Hawaii—just as turtles have swum the seas since the age of dinosaurs. The story on page 34, "Out of the Soup," tells of the trials and tribulations of the endangered reptiles, and of the timely help they're receiving from human allies. (Tom Stack and Associates © Dave Fleetham)

LAUNCHING THE NATURAL ARK

By James R. Udall

"Free the Lobo!"

A petition bearing this slogan recently circulated through the biology department at the University of New Mexico, gathering signatures with ease. Supported by the Mexican Wolf Coalition and the Wolf Action Group, its aim was to introduce Mexican gray wolves bred in captivity to parts of their historic range in New Mexico and Arizona.

After some reflection, James Brown, a biologist best known for his landmark study of mammal extinctions in the Great Basin, decided not to sign. "I'm all for the lobo," Brown explains. "But there are only 39 adults left. Their former habitat is so fragmented that reestablishing a viable population may not be possible. I'm just not sure there's a place for the wolf in the Southwest anymore."

This statement—which, in essence, consigns one of North America's rarest subspecies to imprisonment in zoos—is considered heresy by many local conservationists, who argue that parts of the Southwest are still big and wild enough to sup-port wolves. But many of those who disagree with Brown about the lobo share his larger concerns about the progressive degradation of wildlife habitat in this country. In symposia and scientific journals, in Sierra Club and gun-club meetings, the traditional approach to wildlife conservation is being reevaluated—and found wanting.

"Our emphasis on saving critically endangered species," says conservation biologist Blair Csuti, "has too frequently resulted in crisis management for individual plants and animals."

Although Csuti doesn't advocate making an omelet out of the next clutch of California condor eggs, he is one of many scientists who believe that the U.S. Endangered Species Act, which focuses on life forms on the brink of extinction, must be broadened to include endangered habitats and ecosystems. The logic is simple and pragmatic: Last-ditch efforts to resuscitate a vanishing plant or animal are heroic but expensive; conservation is easier when a species is still common, before most of its habitat has been destroyed.

Some biologists place Yellowstone Park's grizzly population among the "living dead"—those species on the way out. But that prognosis could change if humans work hard to accommodate the big bear. (Art Wolfe)

Conservationists are quick to defend the Endangered Species Act; in the nearly 20 years since its passage it has been one of their most powerful tools. The act has been instrumental in protecting the northern spotted owl and portions of the Northwest's ancient forests, for example, and has restored to healthy numbers once-faltering species such as the bald eagle, peregrine falcon and American alligator. Despite these successes, however, the loss of wildlife habitat continues.

"We've been winning a few battles, but losing the war," admits biologist and Sierra Club staffer Gene Coan. "It's time to broaden the focus."

Before It's Too Late

For biologists and conservationists, "broadening the focus" is a recurring theme. There is, they say, an urgent need to change the mindset underlying our current approach to wildlife conservation. We need to shift the emphasis from next year to next century, continuing to build lifeboats for critically endangered species but also beginning the larger task of relaunching the natural ark.

"Over the past century, we've managed to preserve many parks and wilderness areas, but they were selected largely on the basis of their scenic and recreational importance," says John Hopkins, chair of the Sierra Club's Public Lands Committee. "Now activists must turn to the enormous challenge of preserving lands of biological significance."

> *We need to shift the emphasis from next year to next century, continuing to build lifeboats for critically endangered species but also beginning the larger task of relaunching the natural ark.*

The push to revise current conservation strategies stems in part from the forecast for the next century: The world's population will continue to mount by 250,000 each day. Our forests will be cut down, and then global warming will kick in. Rainfall patterns will shift. Crops will fail. Governments will dither as billions starve, dragging millions of species down with them. Today the Sahel; tomorrow Brazil, Mexico, Kenya.

The scenario is too grim to sugarcoat. The Era of Life's Impoverishment, ecologist Norman Myers calls it. Michael Soule of the University of California at Santa Cruz dedicated his 1986 book, *Conservation Biology,* "to the students who will come after, who will witness the worst, and accomplish the most."

If current trends continue, according to a recent report to the National Science Foundation, "the rate of extinction over the next few decades is likely to rise to at least 1,000 times the normal background rate . . . and will ultimately result in the loss of a quarter or more of the species on earth."

Although the loss of biological diversity will be greatest in the tropics, where most of the planet's animal and plant species reside, our continent will not be spared. Indeed, ongoing habitat destruction makes the loss of thousands of North American species a certainty—unless we fundamentally change the way we use the land.

Defining Our Losses

Biodiversity is a thorny term, most commonly associated with rain forest preservation. Asked to define it, an ecologist will usually say, "It's more than species" before diving into a thicket of complexities. To explore the different layers and levels of biodiversity, I did a little bushwhacking with conservation biologist Larry Harris, author of *The Fragmented Forest,* a blueprint for preserving biodiversity in the heavily logged forests of Oregon.

There are three levels, he explained. The first, "genetic diversity," refers to genes within species. A century ago, biologist C. Hart Merriam identified dozens of populations of grizzly bears: the well-known Plains grizzly populations in California, Arizona, Texas, Oregon and elsewhere. Today all these animals are gone. "Some people will tell you that we really didn't lose anything, because we still have grizzlies in Yellowstone," said Harris, now at the University of Florida. "Nonsense. We lost a lot."

What we lost was genetic information, the DNA coding that enables organisms to cope with change. The bears that vanished were all the same species, but because they had adapted to different environments they weren't all the same. Genetic diversity can also be lost through habitat fragmentation and subsequent inbreeding among a stranded population. This is why the long-term survival of Yellowstone's grizzlies is problematic, and why even captive breeding may be too little, too late to save the Florida panther.

"Species diversity" refers to the variety of species within a habitat or broader area. One place rich in this respect is the 502-square-mile Gray Ranch in southwestern New Mexico. Recently purchased by The Nature Conservancy, the site contains more mammal species than any national park or wildlife refuge in the lower 48.

The Gray Ranch also has an abundance of the third type of diversity biologists look for: "ecosystem diversity." This term refers to the variety of habitats within a region. "One reason the Gray Ranch is so rich in species is that it has so many different kinds of habitats," says Harris.

Wildlife Links

Maintaining a healthy ecosystem requires protecting all three levels—genetic, species and ecosystem diversity. Otherwise, significant losses at one level are liable to produce cascading losses at others. For example, some scientists attribute Yellowstone's small beaver population to the fact that the park's willows—an important food source for beavers—have been overbrowsed by elk. The elk in turn have proliferated because their chief predator, the gray wolf, has been exterminated. (Large predators often function as "keystone" species whose

impacts profoundly affect entire plant and animal communities—one argument for reintroducing them, wherever possible, to habitats from which they've been removed.)

To preserve biodiversity over long periods of time, natural processes must also be protected, including the nutrient, hydrologic and fire cycles that shape ecosystems. "The Everglades are in trouble because of human-caused changes in the water cycle," says Harris. "Within a few decades, all of the romantic elements we associate with that national park—wading birds, red wolves, roseate spoonbills, Florida panthers—may be gone." The park will still exist, of course, but, in the words of conservation biologist Reed Noss, "Scenery is a hollow virtue when ecological integrity has been lost."

What is happening in the Everglades is not unusual; with some exceptions, this continent's ecosystems have been losing natural diversity since the first hunters crossed the Bering Strait. Although many of our landscapes are still scenic, and some are still reasonably healthy, on the whole they have been hollowed out like so many ripe pumpkins by trapping, plowing, logging, mining, damming, poisoning and other forms of human intrusion.

We Still Have a Chance

What if by some quirk of history, the first European settlers had wanted to conserve North America's natural wealth rather than plunder it? If they had known what conservation biologists know today, they would have set aside at least two big blocks of land, four times bigger (say) than Yellowstone Park, in each of the country's bioregions. (Big, to function properly over long periods of time; two blocks, because you don't want all your bears or wolverines or redwoods stranded in one spot.)

Historically, of course, the only place where we had an opportunity to adopt such a farsighted plan in advance of large-scale human impact was Alaska. But in the other 49 states we still have a chance, and a need, to work backward toward that ideal. Though we may never achieve it all, we can achieve a great deal. According to Forest Service ecologist Hal Salwasser, "If biodiversity has a chance anywhere in the world, it is in North America."

But the task is immense. We've been beating back the forest for centuries and it will take an equal effort to recover what we've lost. "The object," says landscape architect Leslie Sauer, "is to gradually shift the impact of our actions from being largely negative to being largely positive."

> *If biodiversity has a chance anywhere in the world, it is in North America.*

One of the first things we must do is reevaluate our existing system of national parks, wilderness areas and wildlife refuges. Because these areas were chosen based on criteria that have little biological relevance, Harris calls them "an agglomeration of artifacts." These wildlands fail to promote biodiversity in two ways. First, regions and habitat types are unequally represented. The West is relatively rich in protected areas; the East, depauperate; the Midwest, bankrupt. We've saved plenty of snowcapped peaks and scree slopes, but lower-elevation forests, wetlands, grasslands and coastal areas are underrepresented or, too often, absent altogether.

Second, many reserves are too small to maintain all the species once found within them. According to biologist James Brown, "Refuges of less than 125,000 acres in which animals are tightly confined may lose more than half their species in a few thousand years." Thirty percent of the United States' national parks and 93 percent of its wildlife refuges fall into that vulnerable category.

The significant qualifier is "tightly confined." Nearly all reserves were originally embedded in a natural matrix; their boundaries were permeable. But now, as population growth and development continue, many parks and refuges are gradually being transformed into islands in a sea of humanity. (Roads are a chief culprit. "The highway system is just a killing machine," says Harris, noting that automobiles are now the leading cause of mortality among Florida's endangered mammals. The sole exception there is the seagoing manatee; its leading cause of death is collision with motorboats.)

Worldwide, habitat fragmentation weighs most heavily on wide-ranging carnivores—the wolves and big cats. But fragmentation also harms large herbivores, primates and bears; habitat specialists like the giant panda and black-footed ferret; and species that require virgin forest, such as the lynx and red-cockaded woodpecker. Small creatures are not immune. As forests in the East shrink under development pressure, even songbirds and plants with modest spatial needs find themselves threatened.

Among wildlife managers there is a growing realization that fragmentation puts at risk everything conservationists thought they had saved. The fallacy of viewing the national park as a kind of ark has been exposed. Warns conservation biologist David Western, "If we can't save nature outside protected areas, not much will survive inside."

Protecting Habitats

To prevent our parks and refuges from slowly losing their natural richness, it is essential to buffer them from development. According to William Penn Mott, former director of the National Park Service, "Stopping our concern at the boundary has got to be replaced with practical applications of buffer zones, regional planning and consideration of the social and economic conditions adjacent to reserves." The benefits and difficulties of implementing these ideas are being demonstrated in the Rockies, where federal and state agencies, prodded by concerned citizens, are attempting to fashion a management strategy for the Greater Yellowstone ecosystem. The same grand experiment needs to be tried elsewhere—in the Everglades, the Cascades, the Sierra Nevada and the Appalachians, for example—to create cohesive landscape units rather than mosaics of bureaucratic turf.

Wildlife corridors and protected lands called "stepping stones" can also help strengthen isolated refuges. By connecting existing parklands, stepping stones can aid migrating birds and wide-ranging predators and herbivores, while helping to link together patchwork landscapes.

"Creating corridors will be of particular importance in the East," says Harris, "where many existing reserves are simply too small to contain even a single panther or a viable population of black bears." An encouraging recent example was the Forest Service's purchase of 29,000 acres in a corridor that will eventually connect the Okefenokee National Wildlife Refuge and the Osceola National Forest along the Florida/Georgia border, a move that will dramatically enhance the ecological integrity of both areas.

Wildlife corridors are not a panacea, however. They help animals more than plants, and some animals more than others. They can also pass along pests and diseases, and their effectiveness decreases as they become narrower. Despite these limitations, corridors are still one of the most effective tools we have for buttressing existing nature preserves.

Another urgent task is to "map the gaps and buy the hot spots"—to locate and purchase rare natural communities that aren't currently protected. In many countries, including China and England, biological inventories are a top priority. Not here. If you want to know how the land is shaped, the U.S. Geological Survey will sell you a topographic map. If you want to know what minerals are underground, fine—they have that mapped, too. But if you want to find out what's on top of the ground, to locate an undisturbed stretch of riparian habitat in Arizona or a rare plant community in California, the government can't help you.

Federal agencies have provided limited support to some private efforts, however. With federal help, The Nature Conservancy has made great strides toward completing an inventory of the nation's rare and endangered species. "Gap analysis," a satellite-mapped program developed by the U.S. Fish and Wildlife Service, has also proven its worth in Idaho and may soon be employed nationwide.

Enlightened Management

Although new buffer zones, wildlife corridors and protected hot spots will help boost existing parks, wildlife refuges and national-forest wilderness, our biological heritage cannot be preserved solely by these means, for such lands cover only 7 percent of the United States. About 18 percent is currently used for agriculture. Cities and pavement claim another 3 percent. Biologist James Brown believes that the biodiversity battle will be won or lost on the remaining 72 percent of our land—the area he calls the seminatural matrix or the semiwild. In the West, much of the semiwild is public land, managed by the Forest Service and Bureau of Land Management (BLM); in the East, it is mostly private. The semiwild now hosts a variety of human activities; the challenge facing conservationists is to balance biodiversity protection with grazing, timber-cutting, recreation and mining. According to the Forest Service's Salwasser, it should be possible to "protect genetic resources, sustain viable populations, perpetuate natural biological communities and maintain a full range of ecological processes while also meeting human needs."

The key will be enlightened management, based on the principles of conservation biology. The Forest Service, BLM and private landowners will have to start, in Aldo Leopold's words, "thinking like a mountain." Whether the "crop" these lands provide is sawlogs, grass or recreational user-days, the objective should be to harvest a sustainable yield without damaging, or further fragmenting, the wildlife habitat.

What about the semiwild and "metroforest" along the densely populated Eastern Seaboard? Although these areas no longer meet the classic definition of wilderness, they still have a role to play in the preservation of biodiversity. Indeed, as the human population of the East continues to grow, the remaining undeveloped lands become even more valuable as wildlife havens. Fortunately, nature is resilient—after an absence of many decades, cougars have returned to New England—and the eastern landscape has not reached the point of no return. Says Zev Naveh, an Israeli ecologist, "My part of the world has been heavily grazed for thousands of years. Compared to the goatscape we are working with, your country offers wonderful opportunities."

---

*F*ortunately, nature is resilient—after an absence of many decades, cougars have returned to New England.

To seize those opportunities, says Leslie Sauer, "We in the East must set our sights higher. There should be no place where there is no wildness at all, no site where natural values are not deemed important. It's time to make a habit of restoration, to take streams out of pipes, rivers out of channels, pavement off of meadows."

Large tracts of wilderness in the East

are rare, but Reed Noss says "wilderness recovery areas" could be created in the Appalachians by closing roads, restoring habitats, allowing natural fires to burn and reintroducing extirpated animals such as the panther, bison and elk. In the Northeast, the Sierra Club and other conservation groups are working to establish extensive reserves within the 26 million acres of forest that carpet Vermont, New Hampshire, Maine and northern New York.

Such ideas may seem visionary. But so was the notion, once upon a time, of even a single national park.

Volunteers Needed

The movement to preserve biodiversity represents a natural evolution of conservation strategy. It is both the next step on the road to a land ethic and the ultimate grassroots challenge. Meeting that challenge will require sweat, dedication and a willingness to take a new look at the American landscape. Nature preserves, new forestry and grazing practices, wildlife corridors, more enlightened land management, public participation at all levels of government, restoration—each will have a role to play.

For those who decide to take on these tasks, the work could be satisfying—just as it was for a handful of Sierra Club volunteers who began trying to restore a degraded park on the outskirts of Chicago 14 years ago. They prowled railroad grades gathering native seeds, cut thickets that were choking ancient oaks, lit fires to kill invading weeds and perused nineteenth-century journals for botanical insights.

As news of their labors spread, more and more people turned out to help. This year [1991] more than 3,000 people, a veritable Noah's army, will volunteer their time and talents. Their mission is to salvage something of tremendous significance, a landscape that was said to exist no longer in North America—a tallgrass savanna, a grassland with trees.

> *T*his year more than 3,000 people, a veritable Noah's army, will volunteer their time and talents. Their mission is to salvage something of tremendous significance.

Nature Conservancy staffer Steve Packard has been one of the project's guiding lights. Asked what his experiences have taught him about preserving biodiversity, he replies, "Restoring the prairie has been like building a cathedral. Thousands of people are involved, and none of us understands more than our own little part of it. Sometimes I visit the site, and the birds are flying around or the flowers are blooming, and I think to myself, 'My goodness, all this depended on us.' And it did. It feels wonderful."

From *Sierra*, September/October 1991. Copyright © James Udall. Reprinted by permission.

Rescuing Rare Beauties

By Doug Harbrecht

So what is it about this bug anyway? For centuries, poets and scholars have been going gaga over Lepidoptera. Even the insects' most eloquent twentieth-century champion, writer-scientist Vladimir Nabokov, caustically observed that his beloved order of insects has been the subject of more "stale anecdotes, pseudo-Indian legends and third-rate poetry," than any other arthropods on the planet. And if truth be told, butterflies do not always behave the way we expect. Many species prefer hanging around rotten fruit, mud puddles, human sweat, even animal scat to flitting among flowers on a sunny day.

But don't tell that to Ro Vaccaro. This very twentieth-century legal secretary happened upon Pacific Grove, California, as the monarchs were gathering by the thousands from all over the West at one of their few U.S. wintering grounds. One dazzling glimpse was enough. In 1988, she sent her resignation back to Washington, D.C., via Federal Express.

Today, Vaccaro is president of Friends of the Monarchs, a 180-member group she started. Last year [1990] she helped convince the town of 17,000 to vote itself a tax hike in order to save a 2.7-acre butterfly resting site from development. That way, she and hundreds of visitors can continue an annual spring ritual: applauding as the monarch males wrestle the females from the sky and then drag them up nearby tree trunks for a roll in the Monterey pines.

And don't badmouth butterflies to Ronald Boender. An electrical engineer who was "retired and bored," Boender decided a few years ago to try butterfly gardening as a hobby in his backyard in Broward County, Florida. "It was a life-changing experience," says Boender, who has since founded Butterfly World in Coconut Creek, a three-acre haven of plants for rare tropical and North American species.

Boender's facility has become one of the largest butterfly research and education centers in the country—a place where hundreds of people come every year for courses in creating butterfly habitat in backyards, highway median strips and corporation headquarters landscapes. "It is so fulfilling," says Boender. "You are helping a delicate creature make the world a more beautiful place."

Back from the Brink

These days, those delicate creatures have a growing number of fans rallying to save them. Scientists, environmentalists and citizen activists are waking up to declines of butterfly species and are pioneering techniques to bring them back from the brink of extinction.

Rare midges, caddis flies and other ecologically important but nondescript insects can vanish unlamented by the general public. But butterflies are a different matter. They have the charisma of pandas and whales, the charm that sparks public interest in an environmental problem. Indeed, butterflies are providing invertebrate conservation with its own poster child.

*B*utterflies are magical. No photograph can possibly give you the sense of what it's like standing in the middle of 50 million monarchs at rest, hearing the faint fluttering of their wings.

"Butterflies are magical," says Melody Allen, executive director of the Xerces Society, an Oregon-based international group dedicated to conserving invertebrates. "No photograph can possibly give you the sense of what it's like standing in the middle of 50 million monarchs at rest, hearing the faint fluttering of their wings."

They're sexy, too. Besides their beauty, the insects' mating habits seem to hold a strong fascination for aficionados: Nabokov, author of *Lolita,* distinguished himself among lepidopterists for skill in identifying lycaenid butterflies. He kept a collection of lycaenid male genitalia, coated in glycerin and meticulously labeled, in vials in his office.

Victims of Urban Sprawl

Unfortunately, butterfly magic of any kind is getting rarer. Colorado entomologist Paul Opler, Xerces Society vice president, witnessed "widespread declines" caused by drought in the last decade. The declines have been leveling off, but tales from local butterfly-watchers have not been promising. Three butterflies that were once found in the San Francisco Bay area are now extinct, the victims of urban invasion of unique habitats in the area. Eight of the 12 butterflies on the federal Endangered Species List could be headed for a similar fate; they are indigenous to rapidly disappearing coastal habitats in California.

Meanwhile, in the Florida Keys, mosquito spraying is unintentionally killing off many of the rare Shaus' swallowtail. And in West Virginia and the Carolinas, entomologists worry about preliminary evidence that southern skippers and other butterflies are highly susceptible to demanol used to control gypsy moths.

Paving, draining and taming habitat plays such havoc with butterflies because many species need specific plants to survive. The great purple hairstreak larvae, for example, eat only mistletoe. Monarchs migrate hundreds of miles to lay eggs on milkweed. And the endangered Dakota skipper is severely limited by its need for unplowed prairie of gama, beard and needle grass. There's hardly any left in the Upper Midwest.

To the Rescue

What to do? First, activists publicize the imperiled species and either entice or attempt to force developers to heed the butterflies' need for untouched habitat. Then, in phase two, conservationists apply techniques of intensive population management.

Consider the case of the El Segundo blue. This endangered butterfly is nothing to write poetry about. Its tiny larvae feed solely on the flower of sea cliff buckwheat once common in the coastal sand dunes that long ago were bull-dozed to make way for Los Angeles's urban sprawl. Now, all but two acres of the dune remnants are found on Los Angeles International Airport property. But in the mid-1970s, when lepidopterists discovered a colony of the insects on a two-acre mound near a refinery owned by Standard Oil of California, the company was happy to protect them. After all, they were butterflies.

Today, Chevron entomologist Richard Arnold supplements older, nonflowering buckwheat on the site with fresh nursery stock. "This butterfly will always be vulnerable to factors over which we have no control," he says, "but you hold your breath and hope for the best."

To the north, the endangered bay checkerspot has been adopted as a mascot by Waste Management, Inc. The company has agreed to deposit $50,000 a year in a trust fund to protect and enhance habitat for a colony of the rare butterflies, which happens to live on Waste Management's property overlooking Kirby Canyon landfill.

The mission blue presents a far more complex—and more common—challenge. For years, scientists were baffled by blues that thrived in some patches of lupines but were absent in others. Now, researchers have discovered that the caterpillars dine on one of the lupine species; adults favor a different one. Another important finding: Ants covet honeydew excretions on the backs of the caterpillars and keep them hidden from predators. Without the ants, the caterpillars make easy prey.

To protect the endangered mission blues, along with the callippe silverspot and the San Bruno elfin, the U.S. Fish and Wildlife Service is trying to reestablish 80 acres of the insects' lupine and sedum habitat in Golden Gate National Recreation Area. The service also en-tered into a controversial agreement with developers on San Bruno Mountain, south of San Francisco. In return for bulldozing some habitat, the developers agreed to protect the remaining portions and set up trust funds for stepped-up habitat "enhancement." To allow this maneuver, Congress altered the Endangered Species Act.

This arrangement is better for the long-term survival of the endangered butterflies, the theory goes, than if no development or management occurred at all. But not everyone agrees. "It's hubris to assume that you can do a better job getting habitat back into shape than saving it," says Robert Michael Pyle, founder of the Xerces Society and former northwest land steward for The Nature Conservancy. "We are far too primitive in our knowledge of many species to reintroduce them later." Pyle hopes Congress will reconsider such agreements when the Endangered Species Act comes up for reauthorization.

Butterfly Dreams

While butterflies have been pinned to boards, classified and reclassified for more than two centuries, ignorance about their behavior is a common lament among conservationists. Under pressure from citizen groups, the city of Santa Barbara and Hyatt Hotels agreed in 1987 to configure a road to a new hotel so that it skirted huge wintering clusters of monarchs. But wouldn't you know it? The next year, the monarchs clustered in trees in the path of the new road, and now the city is trying to renegotiate the agreement.

Some state governments are trying to learn the basics. Ohio has now joined California and New York in funding surveys of threatened butterflies not on the U.S. Endangered Species List, such as Mitchell's satyr and the

Karner blue. And in Washington and Oregon, studies are under way to map locations of all 155 species of native butterflies.

"If you save the insect, you save the habitat. Then you save an entire ecosystem," says Lowell R. Nault, president of the Entomological Society of America. "That's where having your flagship species comes in." If conservationists can help it, people will be dreaming about butterflies for years to come, perhaps creating more third-rate poems along the way.

From *National Wildlife,* August/September 1991. Reprinted by permission.

EARTH CARE ACTION

Out of the Soup

By Jim Watson

Pausing frequently to swat pesky "no-see-ums," Stephanie Richardson kneels over a hole on the beach and scoops out handfuls of coarse white sand. It's a late-summer evening and the sand flies on Keewaydin Island are making her job unbearable. Suddenly, more trouble: About two feet down, the hole starts filling with seawater. Richardson groans. This does not bode well for her "babies."

One night more than two months earlier, a female loggerhead turtle pulled herself onto this beach, dug this nest, deposited 127 leathery, Ping-Pong ball–sized eggs and covered the hole with sand before sinking back into the vast blackness of the Gulf of Mexico. Today, Richardson—a 23-year-old Virginian working as an intern for The Conservancy, a Southwest Florida environmental group—wants to know how many young turtles, if any, emerged to scramble toward the welcoming surf.

The answer hits her right in the nose.

Richardson recoils with a gasp as the stench of death and saltwater rises from the hole. "I love my job, I love my job, I love my job," she chants, leaning back into the task. "I have to keep telling myself that at times like this," she slowly exhumes a pile of drowned, rotting hatchlings and waterlogged eggs, then tallies the carnage: no survivors.

Her luck improves at another nest. Shell fragments indicate that 61 of the 77 eggs laid in this hole hatched. If raccoons or crabs didn't snatch them as they crawled free, the two-inch hatchlings might have reached the open sea, where they stand at least a remote change of surviving.

Turtles' Hurdles

The odds against a single loggerhead hatchling making it to adulthood are astronomical—10,000 to 1, say some biologists. Many eggs never hatch at all. And of the few lucky youngsters that find their way to the ocean, most are quickly gobbled up by seagoing predators.

If nature makes survival difficult for sea turtles, mankind has made it all but impossible. Decades of abuse, ignorance and exploitation have decimated a group of reptiles that has plied the seas since the age of the dinosaurs. Now, more than ever, turtles need all the help they can get.

That's where people like Richardson come in. From Virginia to the Gulf Coast of Florida and beyond, armies of scientists and citizen volunteers are hitting the beach, armed with special state and federal permits. Their mission: to report dead turtles to wildlife officials, care for the injured, monitor nests and, if necessary, move eggs to safer locations.

"These people are our eyes and our ears," says Barbara Schroeder, sea turtle recovery coordinator for Florida's Department of Natural Resources. "Everybody is putting in effort and time to gather the data that will enable us, we hope, to improve the status of sea turtles."

Few imperiled creatures, if any, have ever inspired such an outpouring of public support. "It seems to be something that's growing by epidemic proportions," says Maura Kraus, sea turtle coordinator for southwest Florida's Collier County. "Turtles just do something to people."

A Link
to the Dinosaurs

Why such enthusiasm for a group of uncuddly, wrinkled reptiles—non-ninja turtles, no less? "Maybe it's because they're a link to the dinosaurs," offers Marc Levasseur, who leads turtle walks from the Jupiter Beach Hilton.

Turtle hatchlings that are entering the sea today will not mature for 20 years or more, so it's difficult, if not impossible, to measure the immediate success of these volunteer efforts. Still, the growing wave of public awareness has already borne some encouraging changes.

In 1989, fueled by popular concern and the necessity to protect sea turtles under the Endangered Species Act, the federal government began requiring "turtle excluder devices" (TEDs) on shrimp nets in U.S. waters. (TEDs are basically trapdoor mechanisms that allow the air-breathing reptiles to escape.) And within the past few years, coastal counties and cities, particularly in Florida, have passed aggressive new laws aimed at minimizing hazards to turtles on the beach.

In general, the beneficiaries of this goodwill are loggerheads, the most abundant of the five types of sea turtles inhabiting U.S. waters (there are no comprehensive population estimates). The species is listed as threatened, while hawksbill, Kemp's ridley and leatherback turtles are considered endangered, or nearly extinct. Green turtles are endangered in Florida and threatened elsewhere.

Powerful swimmers capable of traveling more than 40 miles a day, loggerheads sport thick, reddish-brown shells and huge heads (hence their name). From hatchlings no bigger than pocket watches grow adults weighing as much as 450 pounds.

The creatures dwell in warm and temperate waters around the world, venturing farther from the tropics than other turtles to lay their eggs. In this hemisphere, most loggerheads breed along the southeastern coast of the United States, second-largest rookery for the species in the world. Every year, about 28,000 females emerge on U.S. shores from April to September.

"Turtles have been very successful, which

is why they've been around for so long," says Schroeder. "But now people have arrived on the scene, and these animals just can't deal with the things we're doing out there."

Turtles have been very successful, which is why they've been around for so long. But now people have arrived on the scene, and these animals just can't deal with the things we're doing out there.

At sea, as many as 55,000 loggerheads drown in shrimp nets every year, according to the National Research Council. Others choke on plastic trash and swim into boat propellers. On shore, where crucial nest sites are being lost to resorts and condominiums, nesting females can get entangled in beach furniture and crushed by vehicles. Hatchlings emerging from their nests frequently follow the glow of artificial lights into streets and parking lots, where they are crushed by cars or cooked by the sun.

Light, even at low levels, is a powerful beacon for hatchlings. Scientists believe the creatures are "programmed" to crawl toward the relative brightness of the ocean horizon. A light near the beach can skew this guidance mechanism, causing the hatchlings to veer off course. "It doesn't have to be a spotlight," says Schroeder. "It can even be a porch light."

Turtle Protection

People remain the greatest threat to turtles, but people are also the animals' only hope. Leading the charge in the fight to save sea turtles are scientists like Charles LeBuff, a biologist in Sanibel, Florida, who has studied loggerheads for 37 years. In 1968, he started Caretta Research (from the loggerhead's scientific name, *Caretta caretta caretta*), a nonprofit group that organized volunteers to look for turtle nests along the southwest Florida coast. LeBuff has since seen interest in turtles swell. He recalls a workshop he hosted in 1973. "Six people showed up," he says, "and one of them was myself." Recently, a similar conference drew some 500 people from all over the world.

For the past 20 years, Lew Ehrhart, a biologist at the University of Central Florida, has been studying long-term trends in sea-turtle migrations and behavior along a 28-mile strip of shoreline in Brevard County. This Atlantic Coast beach hosts the greatest density of loggerhead nests in the country—20,000 last year [1991].

During nesting season, Ehrhart and his students arise before dawn to research the beach for turtle nests. In the afternoon, they capture immature turtles in coastal lagoons, part of an ongoing study to learn more about the creatures' early lives. By evening, they're back on the beach, clamping tags on adult females.

In the early 1980s, Ehrhart's group and others began sounding alarms about the effects of beach lighting on hatchlings. Newspapers and radio stations carried the message, which was heard by local officials. In 1985, Brevard County became the first to pass an ordinance prohibiting beach lights at night. Since then, 12 Florida counties and 23 cities have passed various lighting laws.

Helping Hands

Because scientists can't cover every nesting beach or keep track of every turtle, they rely heavily on citizen volunteers to supply them with reports from the beachfront. "During nesting season," says Schroeder, "our permit-holders are out seven days a week."

Fortunately, these interspecies Samaritans need little encouragement. "Turtles don't have a PR problem," says biologist Dave Addison. He supervises the sea turtle program for The Conservancy, which has monitored nests near Naples for more than a decade. Although he warns applicants the working conditions are "atrocious," the weather brutal, the pay meager and the insects hungry, he still has to turn people away. Adds Bill Branan, the group's director of environmental protection. "There's a real awareness about the sea turtles and their situation that didn't exist 20 years ago."

Volunteers from all over the country pay to spend a week tagging turtles and swatting mosquitoes on Georgia's Wassaw Island as part of a 20-year-old program at the Savannah Science Museum. "This is one of the few things people can do to get involved with endangered species," says Catherine Blocker, director of education.

---

*T**here's a real awareness about the sea turtles and their situation that didn't exist 20 years ago.***

There's no limit to the lengths (or depths) to which some people will go to help turtles. Take Peter Bandre, a 34-year-old contractor and reptile breeder in Melbourne, Florida. Eight years ago, he founded the Sea Turtle Preservation Society, a group of naturalists that now has 400 members. A few summers ago his turtle love landed him in the sewer.

While investigating a report of disoriented hatchlings in nearby Indialantic, Bandre happened to glance through a sewer grate on the street. Sure enough, there were four or five tiny loggerheads in the muck. He yanked off the grate and jumped to the rescue. He then spied another group in a pipe under a Wendy's parking lot, but this time the grate was cemented shut. He made a call, and soon city workers arrived with crowbars.

As he handed hatchlings up through the grate, Bandre squinted into a jumble of lights and microphones: Some reporters had gathered at the scene. "They interviewed me with my head sticking out of the sewer," he says. Later, the National Marine Fisheries Service presented him with an outstanding achievement award.

Reptile Resuscitation

Few volunteers get as close to their subjects as Beth Libert, known affectionately as the "Turtle Lady" of Florida's Volusia county. Every day for the past nine summers, this 32-year-old former yacht broker from Ponce Inlet has driven her pickup truck along a 13-mile strip of beach near Daytona Beach looking for signs of turtle visits during the previous night. For the first six years she traveled alone, on the job by 6 A.M. Now she gets help from the other volunteers in the Volusia County Turtle Patrol, a group she helped form.

"I'd do anything to save a turtle," says Libert. And she means it. Four years ago, she and a co-worker came upon an adult female loggerhead that had washed ashore. It was partially drowned, probably from an encounter with a shrimp net, and near death. Before an audience of about 60 people who had gathered to watch hatchlings emerge from a nest nearby, the rescuers flipped the beast on its back and began performing CPR. While one volunteer stomped on the turtle's chest, says Libert, "I did alternating breaths through its nose." The animal showed signs of recovery but later died.

What's it like to resuscitate a reptile? "They stink real bad," she says. "They eat horrible things, then spit them up. Their breath smells like rotten clams."

While no one questions the sincerity of these part-time do-gooders, serious scientists like Lew Ehrhart and Barbara Schroeder worry that misplaced emphasis on saving individual turtles obscures the bigger picture. "These people do no harm," says Ehrhart, "but it's important to understand there are larger issues," such as curbing rampant beach development.

For her part, Schroeder is concerned about the noise and confusion that accompany turtlemania. "We don't need any more encouragement for people to come out and comb the beaches looking for turtles," she says. "On Friday and Saturday nights, the beach is a zoo."

"Turtle Lady" Beth Libert would probably agree. People often drive hundreds of miles to watch hatchlings pop out of nests she has been monitoring. When the show runs late, tempers run high. "People have actually picked fights with me," she says. "One guy said, 'I demand that you dig those hatchlings out of the sand right now, or you're paying for my gas!' "

Adds Libert with a sigh, "The turtles I can handle. It's the people that get to me after a while."

From *National Wildlife,* April/May 1992. Reprinted by permission.

EARTH CARE ACTION

A Pitchman for Parrots

By Barbara Nielsen

Sunlight beats down on the weathered tin roofs of Anse La Raye, a seaside village set at the foot of steep green hills on the small Caribbean island of St. Lucia. Along the shore, fishermen are repairing their nets under the shade of coconut palms, while in a sparsely furnished classroom near the center of town, first-grade students, the children of fishermen and banana farmers, are copying spelling words from a blackboard. A typical weekday in this sleepy little town.

Suddenly, a thin Englishman bounds into

Paul Butler communes with a blue and green *jacquot*, the local name for the St. Lucia parrot. (Chris Wille/Rainforest Alliance)

the classroom followed by a five-foot, six-inch "parrot." The first-graders' eyes grow wide, then the youngsters smile shyly as the friendly bird begins flapping his wings, shaking hands and squawking "Hello!" in the native patois.

"This is Jacquot," says the Englishman, cocking his head toward the bird. "He's flown all the way down from the forest to be with you."

Jacquot is the local name for the St. Lucia parrot, the island's national bird, and the children soon learn that he lives nowhere else in the world. What's more, they see slides of his

rain forest home, talk with him about conservation issues (from the need to save the island's forests to keeping the river clean) and giggle as the bird pins parrot buttons on their shirts. The buttons read, "St. Lucia, All My World."

Jacquot's featherless sidekick is Paul Butler, a wacky British naturalist with a knack for making conservation a popular cause. Butler could win a John Lennon look-alike contest. Slightly built, with long, floppy hair, wire-rimmed glasses and a charismatic smile, he is affectionately known throughout St. Lucia as "the parrot man." His efforts have elevated the critically endangered St. Lucia parrot from a slingshot target to a national hero.

The parrot's colorful image now appears on billboards, bumper stickers, T-shirts and St. Lucia passports. A lively combination of classroom visits, reggae songs, music videos, church sermons and puppet shows has made saving Jacquot a cause célèbre with all age groups. As a result, the species' steady slide toward extinction has not only been halted, but its numbers, down to fewer than 150 in the 1970s, have climbed to around 250.

"Ask anyone on the island about the parrot and they'll be able to tell you its name," says Butler, "and many even know the Latin name, *Amazona versicolor*. The people have come to love and cherish the bird as a symbol of their country.

"And because Jacquot talks to them about other conservation issues, they can tell you how deforestation causes erosion, why you shouldn't disturb a coral reef, and so on. For an out-of-the-way island with a poorly educated populace, that's amazing."

The response *is* amazing. But what is most amazing of all is how Butler has almost single-handedly helped the St. Lucians make an about-face from ignorance about wildlife to a passionate concern for an endangered bird—

an achievement that Butler is quick to down-play. He prefers instead to share credit with the many caring and dedicated local people who have aided him in his cause.

Selling Conservation

Butler arrived on the island of St. Lucia in 1977 as part of a task force to study the green-, blue-, and red-plumed St. Lucia parrot. Then a student at Northeast London Polytechnic, he was dismayed to learn that the versicolor pop-ulation had been decimated to a pitiful remnant of the colorful and noisy flocks that had once filled the rain forest with their squawks. Similar problems were widespread in the Caribbean: Hunting, illegal trade and habitat destruction were wiping out populations of many unique bird species—including parrots, macaws and parakeets.

Butler and his colleagues made several recommendations to St. Lucia's Forestry Divi-sion. These included educating the islanders about the need to save the parrot, establishing a reserve to protect the bird's habitat, blazing a trail through the rain forest so rangers could raise much-needed revenue by conducting tours, and greatly increasing the fine for killing or capturing parrots. The law then on the books had been written in 1885 and exacted a penalty of less than $10; worse, it was rarely enforced.

To implement such a plan was a tall order for a tiny country without much money, but Gabriel Charles, then head of the Forestry Di-vision, embraced the recommendations. "And this was before ecology had become a serious issue," Butler adds.

Charles invited Butler to return to St.

Lucia to oversee the program, and the Briton, then only 21 years old, took up the challenge. He was paid a civil servant's salary and lived in a shack without electricity in the rain forest.

A simple, no-frills man who dresses in rumpled khakis, cotton T-shirts and sturdy san-dals, Butler looks back at those bare-bone days with a trace of nostalgia. From the beginning, he was convinced that the best way to reach the people of St. Lucia was to appeal to their sense of national pride. Because his budget was so limited, he was forced to be creative. He began by visiting classrooms and using his offbeat sense of humor and his rapport with children to make the encounters memorable and fun. And by linking the survival of the St. Lucia parrot with the people's sense of national pride, he was able to persuade local businesses to underwrite the costs of buttons, bumper stickers and other promotional materials, thereby connecting their products with a patri-otic cause.

To Kill a Parrot

Within a few years, St. Lucia's Forestry Division had achieved its goals. A parrot re-serve was established, wildlife laws were up-dated and strictly enforced and revenue from rain forest walks had increased the division's budget by a third. But most significant of all was the way the locals had taken up the cause.

*P*eople used to hunt and kill parrots with sling-shots, just for sport. So to see them change their attitudes and rally behind a bird was very heartening.

"Traditionally, people on St. Lucia would kill anything that moved," says Butler. "They used to hunt and kill parrots with slingshots, just for sport. So to see them change their attitudes and rally behind a bird was very heartening."

Over the course of the ten years that Butler spent with the St. Lucia Forestry Division, he gradually fine-tuned his conservation program. By the time he left the division, in 1988, the campaign to save Jacquot and protect his rain forest home had taken on a life of its own, and the staff had turned its attention to other environmental concerns.

Butler's departure was prompted by a challenge from the RARE Center for Tropical Bird Conservation, a Philadelphia-based organization dedicated to preserving the wild plants and animals of Latin American and the Caribbean. (The acronym comes from the former name—Rare Animal Relief Effort.)

"We asked Paul to launch a similar awareness campaign on the nearby island of St. Vincent," says RARE director John Guarnaccia, "but we challenged him to do it in one year instead of ten."

The Root
of the Problem

Butler and RARE were a perfect match because the organization's philosophy, like his, is to get right to the root of every problem. Butler accepted the offer and spent a year on St. Vincent. By the end of his stay, the endangered native parrot, dubbed Vincie, had captured the islander's hearts. Next on the agenda, again with funding from RARE, was a year on the island of Dominica, where the imperial parrot took center stage. In both countries, classroom visits, music videos, bumper stickers, calypso tunes and other catchy sales techniques helped spread the word. As a result of Butler's efforts, conservation laws were updated, wildlife reserves were established and the islanders rallied behind their national birds.

"The approach is similar to a media blitz in which each piece builds upon the rest," says the naturalist-turned-ad-man. "When the children come home from school, they tell their parents what they've learned about Vincie (Siserou on Dominica) and the rain forest. When their parents turn on the television, they see music videos about the parrot and its critical habitat. And when they go to church, they're told about the importance of saving the rain forest. So by the end of the year, they all have a much greater understanding of the endangered bird and its habitat."

Administrators at the RARE center were so impressed with Butler's success on St. Vincent and Dominica that they offered him a full-time position as director of their Caribbean program. Butler accepted but, ever the innovator, suggested that they greatly speed up the process rather than moving slowly around the islands, a year and a country at a time.

"I suggested writing a 'cookbook' with step-by-step instructions," he explains, "and then hiring local cooks to follow the recipes." The result was a 150-page guidebook, *Promoting Protection through Pride.*

The cookbook method was tested on the island of Montserrat in April 1990. The endangered Montserrat oriole was chosen as the focal point, and Rose Willock, a local educator and radio producer who has never even seen an oriole but who worked well with people, was hired as the cook. Within a month, Willock had songs on the radio and bumper stickers in production and was visiting classrooms dressed up as an oriole.

At the time, Montserrat was still reeling from the devastating effects of Hurricane Hugo. The oriole campaign helped to lift the nation's spirits. Children laughed at the sight of a talking bird, and painters and carpenters worked to the happy beat of reggae "oriole" songs.

One Island at a Time

Cheered on by these successes, Butler is now hopping from one Caribbean island to another from his home base on St. Lucia. He is working closely with newly hired cooks in the Cayman Islands and the Bahamas, and he periodically checks on other islands that RARE has helped in the past.

"We're limited only by our funds," says Butler. "The plan is to keep moving through the islands and helping where we can."

Despite his role as master chef, Butler considers the local people the real heroes. Their involvement has ensured that concern for the environment will continue long after he is gone. He looks with pride at St. Lucia, which after a five-year lull is once again pulsing with information on Jacquot. People dressed in parrot cos-

The friendly five-foot, six-inch "parrot" flaps his wings and walks around the school yard, shaking hands and squawking, "Hello! Hello!" (Chris Wille/Rainforest Alliance)

---

*P*aul Butler considers the local people the real heroes. Their involvement has ensured that concern for the environment will continue long after he is gone.

---

tumes are flapping about in schools, the airwaves are filled with songs about Jacquot ("*Amazona versicolor*, de national bird . . ."), children are putting on parrot puppet shows, and ministers are preaching the conservation word.

These approaches have once again been enthusiastically received, but the talk of the island is another of Butler's devices; a "parrot" bus that squawks "Save the forest! Save the forest!" as it putters around. On the outside of

WHAT YOU CAN DO

To find out how you can get more involved in saving endangered species, write: Endangered Species Campaign, Office of Legislative Affairs, National Wildlife Federation, 1400 16th St. NW, Washington, DC 20036-2266.

the bus, a St. Lucia parrot flies through a gaily painted rain forest. Inside, the seats have been removed and exhibits cleverly demonstrate important conservation lessons.

One display shows a pristine forest with a river running through it. When you push a button, rain falls, drains into the river and washes out clean. The next exhibit shows the same scene minus the trees. This time when you push a button, rain falls, drains into the river and emerges a dirty brown.

"The people have heard about deforestation and soil erosion before," says Butler, "but the models are explicit. The people actually see the water coming out brown, and it makes quite an impression, especially on the farmers."

After the bus has visited every church, school and village on St. Lucia, the exhibits will be removed and replaced with a new set—perhaps about litter or coral reefs or mangroves. Butler envisions similar buses chugging around the other islands when funds permit. Meanwhile, bird-costumed humans and reggae songs are getting the word out.

As Butler's Caribbean program shows, dedication and imagination can change people's attitudes and help save the world, one little island at a time.

From *Wildlife Conservation*, January/February 1992. Copyright © Barbara Nielsen. Reprinted by permission.

A Turn for the Better

By Downs Matthews

See them? Little silver and black semicolons punctuating a page of sand?

They're least terns, or *Sterna antillarum*, if you prefer. Dainty, graceful "sea swallows," so called because of their long, pointed wings and deeply forked tails. Satiny white bodies. Black caps. Black-tipped yellow bills, in summer. Silvery gray wings. Fishermen used to call them little strikers, referring to the way they plunge headfirst into the water to catch small fishes and crustaceans. Seabirds brought to near-extinction in the last two decades of the nineteenth century and the first decade of the present one by plume hunters supplying the millinery trade with adornments for women's hats.

Motoring east on Highway 90 through Gulfport, Mississippi, you'll see least terns by the thousands in what has become the world's largest colony of these aerial sprites. Through a happy combination of habitat and human tolerance, they have found a home here.

"Least Tern Area," a sign proclaims. "Nest in Peace."

But only lately have healthy numbers of least terns enjoyed this peaceful nesting haven.

As recently as 1972, Mississippi birders counted just 24 pairs of least terns attempting to nest on a portion of the 26-mile-long man-made beach that lies between Highway 90 and the Gulf of Mexico. Now, some 6,000 pairs of adult birds tend their eggs and chicks here each spring and summer. Clearly, someone helped this endangered species take a turn for the better in Gulfport.

It wasn't Harry Truman. In 1950, when he ordered the U.S. Army Corps of Engineers to build the beach with sand dredged from the bottom of the gulf, President Truman said it would "contribute materially to national defense."

It wasn't the board of supervisors of Harrison County . . . at least, not at first. They hoped a beach would attract tourists eager to pay for fun in the sun, to the economic benefit of impoverished Mississippi.

Instead, the impetus, as it so often does, came from a few concerned citizens for whom the bird rookery was a microcosm of the greater world around them. Judith A. Toups, who writes a weekly column on Mississippi birds and birding for the Biloxi-Gulfport *Sun Herald*; the late Ethel Floyd, who preceded her in that job; and Dr. Jerome A. Jackson, who teaches ornithology at Mississippi State University were among the leaders.

"It was in 1972 that we began trying to get people to give the terns a break," Toups recalls. "It was people versus terns, and the terns were losing."

Almost Swept Away

The main threat, more lethal than storms and predators, was cleaning crews, charged by

When capturing food for its young, an adult tern holds its catch crosswise in its bill and carries it back to the colony. Both parents feed their chicks until the rookie fliers have mastered the art of plunge-diving for fish. (Dan Guravich)

the county supervisors to keep the beach litter-free and pretty for tourists. Routinely, the men would drive their sweeping machines methodically through the tern colony and gather up eggs and chicks along with soda and beer cans and sandwich wrappers. The sweepers had nothing against birds. They were, as they explained, just doing their jobs.

Appalled, Toups, Floyd and other birders tried desperately to protect the birds.

"I would go down to the beach and pound into the sand wooden stakes with little cloth flags tied to them to mark the terns' nests, which are nothing more than shallow scrapes, so the sweepers could avoid them," Toups says. Then the Boston native would go home with blistered hands to care for her sea-captain

husband and six children, wondering if there wasn't a better way.

Finally, the worried conservationists realized they couldn't do it alone. They needed to organize. So Toups and Floyd founded the Mississippi Coast Audubon Society in 1975 and began signing up members. The society's first project, they decided, would be to create a beach sanctuary for least terns.

"I'm not a fire-breathing environmentalist," Toups says. "I know you can't change the world in a day. Environmental causes have to be approached with balance and common sense. I prefer to be a patient plodder who achieves a little at a time."

Toups's first step was to write a letter to the board of supervisors, pointing out the birds' plight and requesting that maintenance crews be instructed to stay out of the rookery during the spring nesting months. Floyd followed up with a petition presented in person to the county fathers. In her newspaper columns, she asked concerned citizens to support the rookery proposal.

Although most of the supervisors were unaware that the rookery even existed and wouldn't have recognized a least tern if one had come to the meeting, they were sympathetic and willing to help. They agreed to set aside two miles of beach as a sanctuary and told the trash collectors to steer clear of it from March through August. They also agreed to help the newly formed society put up signs and fences and notified law enforcement officials that the two-mile stretch would be off-limits to picnickers and bathers.

Of paramount importance, Toups points out, was the fact that the supervisors included the tern colony in Harrison County's master administrative plan. "That made it official," Toups explains, "and not something that would be forgotten about when the newness wore off."

Champions at Work

To this day, Toups remains in awe of what can be accomplished by a few people who take the trouble to try.

Corporate and individual donors lent their support and gave money to the project. Other Audubon societies in Mississippi and the Mississippi Ornithological Society contributed funds, too.

There were complaints, of course. Some people objected to birds "taking over our beach," Toups recalls. But most recognized that the restricted area was rarely used by people

and that 24 miles of beach still remained open for human use.

Others complained about "vicious" birds, when the delicate least terns—ten inches long at most—dive-bombed them in defense of their eggs and chicks. But for the most part, the majority of Harrison County's citizens found it hard to take such complaints seriously.

There were those who refused to give the birds a chance. Rebellious boys ran wild in the rookery, stomping on eggs. In one macabre incident, people had an "Easter egg hunt" and gathered piles of tern eggs. Grown men set their dogs on the tiny chicks. One day, small boys were seen batting helpless chicks like baseballs. Before they could be stopped, they had killed more than 40 baby birds.

To alert the public, Jerry Jackson photographed the carnage and sent his pictures to the media. The coverage tapped a wellspring of concern. When the terns' advocates started selling bumper stickers and T-shirts to pay a patrolman to guard the area, the response was heartwarming. Terns, it seemed, had their champions in Mississippi.

Vigilant citizens began scolding miscreants and reporting vandals. A few arrests helped make the point that Harrison County was serious about protecting its birds. Gradually, public sentiment shifted from indifference to pride in what was rapidly becoming a tern colony of national caliber.

The potential value of the terns' presence has not been lost on Ken Combs, Gulfport's mayor since 1989. After all, 10,000 Californians gather in San Juan Capistrano each year on March 19—Swallows' Day—to celebrate the arrival of cliff swallows, which return there every year after wintering in Argentina. And 5,000 enthusiasts flock to Hinckley, Ohio, on Buzzard Sunday in mid-March to watch 75 to

100 turkey vultures descend on their nesting ledges. With the largest least-tern colony in the world forming each year in April, shouldn't Gulfport seek fame in song and story? What other city in the United States has set aside two miles of popular and valuable waterfront property for birds?

"The tern colony represents three strong pluses for our city," Mayor Combs says. "It is an esthetic asset, an environmental asset and an economic asset. But aside from earning national recognition for Gulfport, the terns place us in the forefront of a trend toward a national environmental conscience. I see our terns as a symbol of the cleaner, healthier world that we hope the 1990s will hold for us."

Business Resumed

Without much fanfare, the word has spread. People now come, if not by the thousands then at least by the score, to watch and enjoy the terns. With binoculars and cameras, they sit on the seawall just a few yards from the nearest scrapes. Others, as I do, set up blinds along the colony's seaside perimeter to better take advantage of the sun.

A row of markers defines a path where people can walk from the highway to the water's edge without coming too close to nests. But the terns object to my passage just the same. Vaulting into the air, they fly nervously over their nests, shrilling alarm cries that sound like Morse code for the letter N: dah dit . . . dah dit . . . repeated over and over. To chicks, the cry is a command to hide and stay hidden until the all-clear is given.

Darting to within six feet of me, a little battler hovers at eye level, chattering in anger, then veers away. Like a tiny torpedo bomber, another tern exposes its opinion of my presence by splattering me with a dollop of excrement. Soon I am decorated all over with the dedicated birder's brown-and-white badges of courage. But once I disappear behind a sheltering barrier of Mississippi-made Leaf O'Flage blind material, the terns resume the pressing business of hatching eggs and rearing chicks.

People now come, if not by the thousands then at least by the score, to watch and enjoy the terns.

The more serious threat comes not from birders, but from the ever-present fish crows that perch in the gnarled and ancient live-oak trees lining the north side of Highway 90. Although the busy highway bars most predators, including dogs, cats, skunks, raccoons, snakes and opossums, it cannot exclude fish crows, gulls and grackles.

Seeing an unguarded nest, a brazen crow ventures across the highway, only to be met midway by two screeching interceptors sent aloft to challenge it. Resembling a pair of Luke Skywalkers confronting Darth Vader, the terns peck at the crow's head and tail, showing off their superior speed and maneuverability as they flutter nimbly around the ungainly bird, screaming insults.

That's enough for the crow. It returns to the oak tree. On a second raid, it comes in low and fast, flying inches above the passing cars and dipping down over the sand. Spotting it in time, the terns drive it away.

On a third foray, the crow soars high over

the colony, floating slowly on the steady gulf breeze. Seeing an unattended nest, it drops like a lump of coal and seizes an egg before the outraged parents can attack. Clenching its meal in its beak, the plunderer slips back across the highway. The parents seem to confer, and one of them settles determinedly onto the remaining egg. Persistent tern parents will lay two or even three clutches if predators or weather frustrate their efforts. Both will share incubation duties for the necessary 20 days or so until the chicks hatch.

Good Parents

Least terns have developed a reproductive strategy different from that of their larger cousins, the royal and sandwich terns. On the Chandeleur Islands, about 30 miles south of Gulfport in the Gulf of Mexico, royals and sandwiches nest wing to wing, lay only one or two eggs and synchronize their clutches so that the whole colony's chicks hatch almost simultaneously. After the chicks grow a bit, their parents leave them in large groups, called creches, supervised by only a few adults, while they go in search of food. Predators are overwhelmed by the sheer number of prey animals and cannot possibly take them all, so many of the young survive despite their vulnerability.

Least terns, on the other hand, lay two or three eggs in shallow scrapes six to ten feet from their neighbors' nests. The thimble-sized eggs, pale greenish-gray speckled with brown and black, blend into the sand-and-gravel beach. The precocious chicks hatch days apart into fuzzy Ping-Pong balls with legs. They depend on concealment for protection, huddling close to the sand and nosing down beside stray pieces of seaweed and other flotsam while their

parents distract and drive off predators. The terns are good parents. During the 18 days before their chicks can fly, they not only aggressively protect them from predators but brood them faithfully against rain and the heat of the sun.

A tern mother stands over her two chicks with wings akimbo to shade them from the intense sun. Their father, meanwhile, goes fishing. He works beyond the tidal zone, cruising head down, watching, until he spots a silversides or a mullet minnow just below the surface of the bottle-green water. From a height of 20 feet, he dives, and misses. Up again. Search. Dive. Splash. Another miss. And once again, up, search and dive.

Success. The tern comes up with a wriggler. He chews on it to break its bones and returns with it to his scrape. Greeting his mate, he offers the limp fish to her. She declines and steps away from the chicks, who gape eagerly for a meal. He pokes the fish into one baby's gullet, and after watching it disappear, gets back to work. To catch one fish, he has to make, on the average, five attempts. When the chicks fledge, he will have to teach them to fish, a skill tha must be learned over a period of weeks of daily practice. The chicks will continue to rely on their parents for food until they become competent fishermen themselves.

Sharing Space

As August approaches and the nesting season comes to a close, the least tern families leave Gulfport to the tourists and migrate south. In stages, they will make their way to Central and South America to beaches that are safe from the chills of winter.

During the birds' absence, Harrison

County beach supervisor Tim Anderson will muster his maintenance crews to sweep away accumulated debris and plow up sandbur clumps that infest the beach area. Least terns require clean sand or gravel, preferably free of vegetation. And Anderson doesn't want to disappoint them.

Meanwhile, Gerry Morgan—a Biloxi, Mississippi, resident for the past 30 years and a birdwatcher since she was ten years old—and her fellow members of the Nest in Peace Campaign collect donations to buy signs to protect the least tern rookery.

"So many people are curious about the terns," Morgan says. "We want them to enjoy the experience of seeing these wonderful birds up close and to understand how important it is to share our space with them."

From *Wildlife Conservation,* January/February 1992. Copyright © Downs Matthews. Reprinted by permission.

EARTH CARE ACTION

Return of the Taiwan Ten

By Dwight Holing

There was a shared sense of purpose in the Republic of China's decision to release the "Taiwan Ten." Everyone seemed to be affected. China Airlines agreed to fly the ten to Indonesia for free. The Taipei Education Department sponsored an essay contest with the prize being a chance to accompany the ten home. Nine students won and another 98 bought tickets so they could go, too. And when the plane departed for Jakarta, a crowd of dignitaries, well-wishers and reporters extended a hero's send-off.

Who are the "Taiwan Ten"? Political refugees? Former hostages? Actually, they're orangutans, babies that were rescued from the hands of a ruthless gang of animal smugglers. Their return to Indonesia signals a new era of international cooperation in the fight against the illegal pet trade.

Orangutans are among the most endangered animals on earth; they are also among the most sought after by animal smugglers. The going price for a black market baby orangutan is U.S. $5,000. Certified animals command prices of upwards of $100,000. Smuggling, along with destruction of the orangutan's native forest habitat on the islands of Sumatra and Borneo, has reduced the number of wild orangutans to dangerously low levels.

"There are probably only 30,000 left," says Dr. Birute Galdikas. She should know. Galdikas is the world's foremost authority on orangutans. One of the three famous "Leakey Ladies," she has been studying orangutans for more than 20 years. Like Jane Goodall and the late Dian Fossey, the scientist has dedicated her life to the plight of the great apes.

From Mother's Arms

While no one knows how many orangutans are captured and sold each year, Galdikas says pet stores in Taipei prove the problem is of epidemic proportions. A group she heads, the Orangutan Foundation International, recently conducted an investigation in Taiwan. It discovered approximately 700 orangutans. Noting that the typical method of capturing babies is to shoot the mother, Galdikas says each live orangutan delivery represents seven or eight others that died during capture or in transit. "Thousands of wild orangutans have died being smuggled," she says.

Taiwan isn't the only country where orangutans are sold. Alert customs officials at the Bangkok airport last year discovered three wooden crates destined for Yugoslavia. Their suspicions aroused, they opened the crates and found six baby orangutans crammed inside.

Like the "Taiwan Ten," the "Bangkok Six" became the subject of delicate diplomatic maneuvering between the governments of Thailand and Indonesia. The two countries eventually reached an accord that will make it easier to repatriate captured orangutans in the future. Says Indonesia's ambassador to Thailand, Gatot Suwardi, "This act of cooperation between our two countries can only work to help the process of saving all species on this planet."

*E*ach live orangutan delivery represents seven or eight others that died during capture or in transit.

Back to the Wild

Like other captive orangutans that have been intercepted, the "Bangkok Six" were flown to Kalimantan province on the island of Borneo, where Galdikas oversees a rehabilitation station. So far 60 former pets have been successfully reintroduced to the wild. Another 30 are currently in transition. A similar camp operates on the island of Sumatra.

"The biggest success of the rehabilitation program is the support it has won from the local people," says Galdikas. "They now show intense loyalty to orangutans. Before, if an orangutan was eating their crops, they might have killed it. Now they go to their village chief and ask for help."

Galdikas fears this important change in attitude might be wasted if an end isn't put to the illegal pet trade. Public education and cooperation among governments, wildlife conservation groups and the private sector are helping, but the war against smugglers is far from over.

Says Galdikas, "The irony is, the harder

we work and the more effective we are, the higher we drive up the price for illegal orangutans. The economic disparity between Indonesia and countries like Taiwan that can afford to pay $5,000 for orangutans as pets underlies the problem. Until we solve that, smuggling will continue."

EARTH CARE ACTION

A Poacher's Worst Nightmare

By Jessica Speart

Dave Hall doesn't like to lose. And in a career spanning more than 20 years as a star undercover agent with the U.S. Fish and Wildlife Service, he's rarely had to. Originally trained as a biologist, Hall took one look at what was happening to North American wildlife and tossed aside the idea of a research career. He signed up for what he thought would do the most good—law enforcement. It's been a madcap ride ever since.

In 1969, the Fish and Wildlife Service (FWS) sent agent Hall to enforce game laws in New Orleans, the national center for the flagrant poaching of alligators, which were sliding rapidly toward extinction. Hall soon learned that in the bayous of Louisiana, he was the enemy, and outlaw poachers were the heroes. And nobody, least of all some punk game warden, was going to take away their God-given right to hunt all the game they wanted.

In this century, wildlife poaching, particularly of endangered and threatened species, has turned into big business—the FWS estimates the U.S. trade at $200 million annually. A bald eagle feather headdress on the black market can bring more than $10,000. A grizzly bear commands $15,000 as a trophy and $5,000 for an ounce of gallbladder sold for medicinal purposes in Asia. Live peregrines shipped to falconers in Saudi Arabia garner as much as $135,000 apiece. Curbing the trade is not easy—ask one of the nation's most successful, and most controversial, agents: Dave Hall.

When he first confronted the poachers, Louisiana's alligator population had dropped

Pictured here with a walrus tusk,
U.S. Fish and Wildlife Service agent
Dave Hall posed as an ivory dealer
to infiltrate an Alaskan poaching
ring. (Ken Glaser/Black Star)

90 percent from its historic numbers, at least partly because of the skin trade. All alligator hunting had been banned in the state since 1964. But laws didn't interest poachers one bit.

The Gator Caper

With no undercover training, Hall faced his first showdown alone in 1970 when an informant gave him the name of Markita Tompkins Percle of Morgan City. The only woman dealing in gator skins in Louisiana, Percle was notorious. "She was shrewd and didn't care who the hell she cheated," says Hall. "Little bitty damn thing, but you could tell she was mean."

Hall spent six months pretending to be a buyer of skins and earning her trust. He chuckles at the memories: "She used to say, 'There ain't no game warden smart enough to catch me. Who in the hell would think a little ole woman like me with five snotty-nosed kids in a pickup truck would be hauling gator skins?' And I'm thinking, 'Yeah. You wait.'"

Eventually Hall decided to arrest Percle during the first deal when skins and money would be exchanged. Percle was busted with 36 alligator skins and faced a fine of $600 and 90 days in jail. But she walked off scot-free (Hall suspected the case had been fixed), and the informant disappeared, never to be found. The laws against killing gators didn't mean a thing. Dave Hall's war was on.

Tracking Outlaws

Hall went after a man who describes himself as "the most famous outlaw" in southeastern Louisiana. "Nobody was tougher or meaner than Clinton Dufrene," says Hall. "What Clinton liked to do best was to lead game wardens on Keystone Cop chases." Working nights out of an airboat with a powerful aircraft engine, he would outrun the state wardens, leading them onto tree stumps that would send their boats flying like a bayou version of Smokey and the Bandit.

---

*W*ho in the hell would think a little ole woman like me with five snotty-nosed kids in a pickup truck would be hauling gator skins?

Dufrene, today a legitimate and wealthy businessman, says, "I knew Dave was after me real bad. And I loved it. Believe you me, I loved it. That was my favorite pastime. For Dave, I was like the elephant with that long tusk."

Finally Hall showed up (with reinforcements) just as Dufrene was walking out the door of his house, an outlawed shocking machine in hand, apparently ready to do a little illegal fishing. But Dufrene was in jail just long enough to be bonded out by his lawyer.

By this point, Dave was getting mad. He was busting poachers left and right, but they were walking out of jail before he'd even gotten back to his office. So Hall did what he does best—told his story to anybody who would listen. There proved to be plenty of willing ears

at newspapers, television stations and a federal grand jury. Hall and two other agents testified before Congress about the need for more personnel and funding. In the floodlight of publicity, fixing cases for poachers became more difficult. Hall was chipping away, chipping away at the industry.

In the 1970s, gators were federally protected (under an amendment to the Lacey Act and later under the Endangered Species Act), and poachers faced up to $20,000 in fines plus prison terms. The first person to do time was Clinton Dufrene: 22 months in federal prison for two violations. Hall cultivated Dufrene as an informer and picked up the trail of the two largest alligator-skin buyers in the United States, Jacques Klapisch and his right-hand man, Jack Kelly. "Klapisch was quiet, but he was evil," recalls Hall. "Kelly's just a hoodlum, a straight-up hoodlum.

"I had been working on this for years to get to the Top Gun," says Hall. "Everything had come into place." In 1974, a procession of agent cars and a spotter plane—one agent called it a "circus"—traced a certain station wagon up the interstate toward New York. When the driver reached a warehouse in Elizabeth, New Jersey, agents moved in. They found a huge shipment of skins spread out around Klapisch and Kelly.

Klapisch and Kelly each received suspended sentences and $9,500 in fines. But the lesson didn't work. Eighteen months later, the poachers were caught again. This time, they pulled out of the business.

Conservationist of the Year

With the major poachers gone by 1977, Louisiana's alligator population recovered, and

Hall went to bat for a legal season. All along he had been preaching to the poachers that their best hope was to obey the law and let gators thrive. He was right. Louisiana now has so many gators that the state allots a number to be hunted legally every September.

Hall's bulldog determination to rescue the alligator from oblivion and protect the region's wildlife earned him the Louisiana Wildlife Federation's Conservationist of the Year award— three times. United States magistrate Michaelle Wynne in Louisiana calls him "one of the best things to hit this country. He was the beginning of wildlife enforcement in Louisiana."

Hall's next major undercover caper began in 1980 when Soviets complained to U.S. authorities about thousands of headless Alaskan walrus decorating Siberian beaches. The walrus had been killed for one reason—ivory. The FWS's response was to patrol the Alaskan coast. Hall's response was, "What the hell good is that going to do?" When challenged to come up with a better idea, he did.

Hall's first step was to take on a partner, William Vaughn Doak, sometime actor and former owner of a nightclub with "seminude dancing," ex-skinflick maker and currently owner of a shop in New Orleans called "Endangered Species." Doak sold only animal artifacts that could be legally traded, and his shop was loaded with antique ivory pieces. He taught Hall what they hoped would be enough to make him convincing, and Hall transformed into Dave Hayes, Texas oil man and aspiring ivory merchant.

"I always tried to go after the big fish more than the little bitty ones," says Hall. This fish was Charles McAlpine, the biggest ivory dealer in the country, who happened to live in Washington, conveniently close to his market in Alaska.

Under the Marine Mammal Protection Act of 1972, native peoples are allowed to hunt walrus and sell ivory carvings. However, selling raw ivory to a nonnative is a federal offense. McAlpine deftly solved this problem by using natives to do his buying and transporting for him. As one native bragged to Hall, "I could take that ivory and drag it around on a string in front of God and everybody, and there ain't nothing they could do to me."

With the shop as their cover, Doak and Hall began buying big time from McAlpine. The two eventually went on spending sprees in Alaska, spreading their net over smaller dealers. Hall had no problem finding quarry. "Everyone was beating down our door, from bartenders to taxi drivers."

10,000 Pounds of Ivory

Among the more interesting characters they investigated were Jerry Kingery and Douglas O'Neill. Kingery, a former Hell's Angel who had formed his own biker group in Anchorage called the Brothers, used a human skull—he claimed it was his ex-girlfriend's— as a table decoration. O'Neill was a former biologist for the state Department of Fish and Game. Wealthy and paranoid, he hired a detective to go to New Orleans and check out this "Dave Hayes." The cover held.

In February 1982, FWS agents closed the net. Coordinated raids in Anchorage, Nome, Seattle, Hawaii, San Francisco and New Jersey produced 100 arrests and 10,000 pounds of illegal ivory. Hall smiles at the memory: "We showed them how the cow ate the cabbage."

Hall received a letter from the FWS deputy director saying the case would go down in the annals of wildlife enforcement. Nothing could ever match it again. But as much as Hall has been praised, he's also been criticized for his arrogance and ego. One fellow agent calls him "the type of guy that, if you're out on a raid, will leap from the helicopter when it's 15 feet up in the air to be the first one charging through the door." Adam O'Hara, head of FWS Special Operations, simply states, "He is a man possessed."

But one man alone can't protect America's wildlife. With a high-rolling international market for rare animals, poachers have many ways to make fortunes. Yet FWS fields only 204 agents to fight this well-equipped enemy. Lack of federal funds ties many agents to their desks

for six months of the year. "There are some days I think I'd take the odds that Custer had," says agent Terry Grosz.

Ultimately, controlling the wildlife trade may depend on the courts. "That's been our biggest problem," concedes Mike Sutton of the nonprofit antipoaching group TRAFFIC U.S.A. "Most prosecutors don't have much interest in wildlife cases, and judges don't take it very seriously compared to a bank robbery or murder."

In the meantime, Dave Hall continues his job of uncovering those who trade in endangered wildlife. "It's gonna be damn hard to win, but being the fighter I am, I'll be one of the last standing."

From *National Wildlife*, April/May, 1992. Reprinted by permission.

Operation Boogar Man

The date: March 3 [1991]. The location: Kuwait City Zoo, three days after Allied troops had entered the city. The scene: rubble, smoke, stench and death.

Near the entrance, Asian macaques wandered about freely, one of them clumsily trailing the others on the remains of a leg that had been shot away. Wounded in the shoulder by a bullet from an Iraqi rifle, an Indian elephant stood silently in its compound. Nearby, two emaciated hippopotamuses sat in a pool of filthy water, barely breathing.

Wounded in the shoulder by a bullet from an Iraqi rifle, an Indian elephant stood silently in its compound.

"It was a sight I'll never forget," says Colonel Philip Alm, the first American soldier to enter the zoo. A veterinarian with 30 years of service in the U.S. Army, Alm was sent into the city immediately after the initial liberation to assess the availability of food and water for Kuwaiti citizens. Though visiting the zoo was not a part of his assigned duties, the 52-year-old military man knew the situation there was critical. "I had to take action," he says. "The animals were dying." In the next few days, he and volunteers under his command worked feverishly to save the remaining zoo animals, as well as hundreds of cows and other domestic creatures in the city.

In a country that abounds with stories of Iraqi atrocities, the decimation of the zoo wildlife and other animals is but one more example of the needless destruction that took place during the seven-month occupation of Kuwait. "I've been all over the world but have never seen anything as bad as that zoo," says John Walsh of the World Society for the Protection of Animals, who visited the site last March.

Silence Shattered

Before the war, the zoo's collection included more than 400 animals, representing 134 species. When Colonel Alm arrived, fewer than two dozen survived. "Some Iraqi troops were billeted at the zoo and apparently were completely undisciplined," he says. "Their officers were living downtown in buildings and hotels and had turned the enlisted men loose. The soldiers had to go out and scavenge food for themselves."

At the zoo, the Iraqis took target practice, shooting monkeys for sport, and killing gazelles, oryxes, birds, even anteaters for food. On many mornings, recalls Abdul Al-Omani, whose house sits adjacent to the zoo, the silence was shattered by the sound of automatic weapons as soldiers slaughtered creatures next door.

Two brothers, Ali and Suliman Al-Houti, who had worked for the Kuwait City sanitation department, did their best during the occupation to care for the zoo animals. "The first time I tried to go there, the Iraqis punched me and stole my watch," says Ali. Finally, they allowed him to push a donkey, which he bought with borrowed money, into the lion's cage. The cats, on the verge of starvation, quickly devoured the sacrificial animal.

Animal First Aid

Colonel Alm spent his first two days in the city searching through neighborhoods for cows and other animals. "Rather than deal with bureaucrats," he says, "I thought I would go to the people to find out what the food situation was. I drove around, asking, 'Where's the boogar?'" (A slang term for the Arabic word *bakara*, meaning "more than one cow.") He soon became known as "the boogar man."

Initially, Colonel Alm's biggest problem was finding a source of water. "If I didn't do something, all of those animals were going to die," he says. "So I did something."

What he did was commandeer an Army tank truck that had arrived from Saudi Arabia with a load of bathing water. It wasn't until later that the veterinarian learned that the water was headed for the camp of some U.S. soldiers, including a brigadier general and his staff. "He wasn't too happy at first," says Colonel Alm, "until he realized why I took it."

Soon other military veterinarians arrived in the city from bases in Saudi Arabia and began patching up the injured animals and cleaning up the zoo. "It was certainly an unexpected rescue operation," says Colonel Alm, "and it provided me with a tremendous amount of satisfaction. But I guess I'll always wonder how people could be so needlessly destructive and barbaric."

From *National Wildlife*, August/September 1991. Reprinted by permission.

Good Hope for Cape's Endangered Medicinal Plants

By Kate De Selincourt

In a bid to protect the rich plant life of South Africa's Cape, botanists and local government officials have joined forces with local witch doctors, or *sangomas*, to organize nurseries for medicinal plants.

The flora of the Cape forms one of the world's six biomes. It occupies just 1.5 percent of Africa but is extremely rich in species.

Rapid urbanization in South Africa is bringing thousands of country people to Cape Town each month. The new arrivals bring with them the tradition of visiting sangomas who prescribe herbal medicines.

Gathering herbs from the wild has become a boom industry, and there are fears that some plants, especially those dug up for their roots or bulbs, may become extinct. Attempts by the police to stop people collecting plants have failed. "Many of the collectors are young men and very fit," says Cameron Greene of Stelenbosch University, who has studied the herb gatherers. "The police chase them through the mountains on motorbikes but never catch them."

On one occasion six sangomas were arrested while collecting bark in a forest on Table Mountain. Fiona Archer, an ethnobotanist at the University of Cape Town, interceded on their behalf, pointing out to the magistrate that

if these collectors were locked up, others would simply take their place.

> *M**any of the collectors are young men and very fit. The police chase them through the mountains on motorbikes but never catch them.*

"I said we would just drive the whole thing underground, and instead we should take this opportunity to cooperate with the healers in finding more sustainable sources for healing plants," Archer explained.

The sangomas were released, and the Western Cape Traditional Plant Use Committee was set up. This committee has now discussed with the sangomas plans for cultivating traditional herbs. The healers are enthusiastic about the idea because it will save them a lot of traveling and ensure them a steady supply of plants.

The committee, chaired by Cape Town City Council's director of parks and forests, Peter Rist, has applied to the South African

Nature Foundation for funding for a full-time worker and cash to start a nursery.

Which Species to Save?

The scheme has the support of many of the local botanists and ecologists—but not all. Ian MacDonald, of the Percy Fitzpatrick Institute of African Ornithology at the University of Cape Town, argues that the scheme gives credibility to medicinal practices which, he says, ought to be phased out, "the way my culture no longer uses eye of newt and wing of bat. What will the position of conservationists be if, five years later, it turns out that they have been encouraging the cultivation of a harmful plant?"

Wouter van Varmelo, spokesman for the committee, agreed that the authorities would need to be careful about which species were cultivated, and they were still discussing how much control there would be over the nurseries. "But if people need the plants for their medicine, we must let them grow them."

The crops will be valuable not only to sangomas, who can sell them in the same way they now sell wild plants. They will also form a reservoir of potential pharmaceuticals. Research is needed to find the basis of most of the traditional remedies before the sangoma's knowledge disappears.

This first appeared in *New Scientist*, London, the weekly review of science and technology, in January 1992. Reprinted by permission.

Snowy egrets, as well as other wildlife, are attracted to a filter bed in this natural sewage treatment plant in Denham Springs, Louisiana. See "Plants That Purify" (opposite page) for more on how scientists are finding new, environmentally intelligent ways to deal with society's wastes. (Neil Johnson)

PLANTS THAT PURIFY

By Christopher Hallowell

Under a warm morning sun in Benton, Louisiana, Carl Janzen bends over to pluck a stem of blossoms from a clump of lush arrowheads. The tiny flowers are a study in simplicity—three sparkling white petals curving out of a brilliant yellow button of pollen-heavy stamens.

Janzen looks down at the plant that bore them, its wide leaves covering the gravel at his feet. "Now this here plant is a good ammonia worker, but you got to watch him or he'll put his roots down too deep and clog up the gravel," he explains in a raspy drawl.

Janzen supervises the sewage treatment plant for the town of Benton, population 2,500. Sprouting from the gravel bed on which he stands are over 3,000 plants. The sewage that flows around their roots and between the chunks of gravel provides nourishment for extraordinary growth. And in the process the gray effluent turns almost crystal clear.

Pointing to a sample of clear liquid end product in a laboratory beaker, Janzen observes, "This here is all going back to Mamma Nature from the word go." His easygoing manner belies the significance of his statement: The citizens of Benton receive their drinking water from the nearby Red River, about three miles downstream from where the treated wastewater is discharged.

Sewage treatment plants exist throughout the country, but in this one plants do virtually all the cleansing. Since it was constructed in 1987, Benton's treatment system has proven to be so efficient that it has become something of a showpiece for the town.

The main attraction, called a "microbial rock-reed filter," is an example of natural sewage treatment technology that is beginning to proliferate in the South and to creep northward. Natural treatment is still viewed warily by some scientists and by engineers who design conventional sewage treatment facilities. Yet in early 1992, about 300 natural systems had been built, according to Robert Kadlec, a chemical engineer at the University of Michigan who has been researching them for over 20

61

At the natural waste treatment plant in Denham Springs, Louisiana, the effluent flows from two ponds over and through the rock-reed filters, after which it runs into an ultraviolet disinfection chamber and then out to the Amite River. (Neil Johnson)

years. Some are as small as the 12-acre one in Benton, but others are mammoth, 1,000-acre-plus constructed wetlands.

How It Works

Benton's rock-reed filtration works on the same principles as any natural system, and it goes like this: About 200,000 gallons of raw sewage a day are pumped into a ten-acre oxidation pond. A floating, wind-driven aeration device anchored in the pond's center churns and oxygenates the resulting soup, allowing aerobic bacteria to slowly break down the solids, which then settle to the bottom as sludge. In this airless

gunk, anaerobic bacteria continue the decomposition process.

The time that bacteria are given to consume nutrients is one of the main differences between a natural and a mechanical sewage treatment system. Benton's sewage remains in its settling pond for three months, ample time for bacteria to chew on and break down pollutants. An average conventional plant flushes sewage through the system within 48 hours or less, not enough time for bacteria to thoroughly reduce pollutants in the solids.

Benton's partially treated wastewater then is drawn by gravity from the pond into the rock-reed filter, a 1,200-foot-long gravel bed that wraps around two sides of the pond. Rows of nutrient-absorbing plants such as African calla lilies, water irises, arrowheads and miniature and giant bulrushes grow in the waterlogged gravel. A small group of plants with large leaf surfaces, such as the African calla lily, can suck up water at a prodigious rate—about 1,000 gallons per day, depending on sunlight—and release it into the air through evapotranspiration. The wastewater spends a month trickling through the rocks and roots of the filter, where bacteria and microbes attached to plant roots further break down pollutants.

The Natural Advantage

Visitors to these natural treatment plants will discover a few surprises: First of all, there is no unpleasant odor. In fact, the systems emit no particular smell at all. Second, some of the facilities have become impromptu wildlife refuges; others double as recreational areas.

There are limitations to natural sewage treatment, however, and one is the large expanses of land required. This may be one reason the South has been the principal laboratory for natural treatment technology. Also, the systems depend on the continuous biological activity of plants and microbes, which slows greatly during the winter months in colder climates.

"I am a real fan of this technology," says Robert Bastian, a scientist with the Environmental Protection Agency (EPA) who has followed the progress of natural sewage treatment since the 1970s, "but I am a little suspicious of the promotion it has been given recently. There's a trade-off required to get clean water. You're going to have to pay for all that land. Also, these systems manage natural processes, and not that much is yet known about them."

New York City produces enough goo every day to cover a football field to a depth of ten feet.

Yet in some instances small-scale natural sewage treatment facilities—which principally consist of simple ponds and marshes—avoid certain problems endemic to conventional plants. They do not require the tons of concrete, the miles of iron pipe, the mechanical equipment and the heavy doses of chemicals. Nor do they suffer the inevitable breakdowns.

They also help reduce the bane of all conventional plants: the quantities of sludge, often including heavy metals, that settle out of the liquid and must be incinerated or hauled off to overflowing landfills—both controversial practices. New York City produces enough of this goo every day, for example, to cover a football field to a depth of ten feet.

The 16,000 plants in this country treat almost 30 billion gallons of raw sewage

WHAT YOU CAN DO

Think before you toss! Remnants of wallpaper, fabric from drapes or upholstery and excess carpeting can often be used for lining drawers, closets and dressers.

daily, and many mechanical plants still discharge pollutants into rivers and oceans. Prime among these are nutrients (ammonia, phosphates and nitrates) that can contribute to the growth of algae—in blooms or in broad surface mats—that suck oxygen out of the water, decimating certain forms of aquatic life. Some heavy metals, toxic chemicals and pesticides also make their way through conventional systems.

The Mighty Microbe

The key processes of natural sewage treatment are entirely invisible. Plant roots nurture and shelter countless microorganisms that not only consume ammonia, nitrogen and phosphorus but also attack and break down such pollutants as industrial chemicals, detergents and pesticides into simple compounds that the plants can then absorb.

How do they do it? "Each plant secretes certain chemicals that allow specific microbes to live in harmony with it," explains B. C. Wolverton, Ph.D., a pioneer in natural wastewater treatment. "And in some cases microbes mutate in response to certain chemical variations and become even more efficient at digesting toxins."

Wolverton, a microbiologist and environmental engineer, has been researching the ability of plants to clean polluted water for more than 20 years. He spent much of that time at NASA's Stennis Space Center in Picayune, Mississippi, developing a way to recycle waste at the space agency's planned moon base. While doing so, he devised a waste treatment system for the space center that uses the water hyacinth as a cleaning agent. Now retired, he heads Wolverton Environmental Services, which advises towns and cities on natural treat-

ment systems. Rock-reed filters like the one at Benton are his design, though the basic idea has long been used in European countries, principally Germany and the United Kingdom.

The Unpredictable Nature of Nature

In spite of Wolverton's work and that of other researchers, wariness toward natural waste treatment technology still exists. Waste management engineers who are trained in traditional methods are uncomfortable with the idea of depending too much on the vagaries of nature and the shifting relationships among pollutants, plants and microorganisms. Their skepticism has impeded the general acceptance of natural treatment systems.

"These systems are full of unknown," says Mike Giggey, a project manager for Wright-Pierce, an engineering firm that designs municipal waste treatment facilities in New England. "There's no doubt that constructed wetlands have the potential for improving water quality, but it's very difficult at this point for an engineer to know how big to make the wetlands or what kind of vegetation to plant. Getting from the research arena to the applied situation takes a long time."

Cost Considerations

Many towns are considering natural treatment plants out of purely financial concern. Take the town of Picayune, population 12,500. In the late 1960s the city put in a conventional waste system, which is now falling apart. After a typical Gulf Coast downpour one day last July [1991], the system was full of odoriferous sewage that spilled out of its reservoir in torrents, mixed with street runoff and flooded into the Pearl River, just upstream from a state wildlife refuge.

Faced with a repair bill of $11 million, and no help from the Bush administration, the Board of Aldermen recently voted to deactivate the antiquated plant and replace it with a natural system. The estimated cost: $350,000. "As an amateur environmentalist, what I like about the natural systems I've seen is that they harbor turtles," said Jim Young, Picayune's city manager, as he viewed the bursting old plant. "That, to me, means something is happening right."

Birds and Flowers Like It, Too

Young puts his finger on another benefit of natural waste treatment systems: They can be pockets of floral and faunal richness, of no small importance as wetlands vanish at the astounding rate of nearly 300,000 acres per year.

On early summer mornings, purple martins and swallows sweep through the shrouds of mist rising from the two oxidation ponds of the Denham Springs, Louisiana, rock-reed filter plant, which serves 20,000 people. A forest of giant bulrushes grows in the 15-acre rock marsh into which sewage from the ponds flows. One morning a five-foot alligator lay partially concealed in the grass on one of the pond's levees, its glistening eyes the only hint of life within the still shape. A colony of red-winged blackbirds perched on the long stalks, fill-

ing the air with clucking and raspy beeping that almost drowned out the whine of trucks on nearby Interstate 12.

"Even though the water may not be fully treated, these sites are receiving a lot of birdlife and are ideal nesting sites," says Lynn McAllister, a biologist studying the effects of constructed wetlands on bird and animal life for the EPA.

Two of the largest constructed wetlands in the country cover 1,200 acres each; they serve Orlando and Lakeland, Florida. Both attract wildlife and contain deep pools where aquatic life flourishes. Orlando's Wilderness Park is also home to rare white pelicans and several pairs of bald eagles. Over 100 pairs of wood storks nest in Lakeland's artificial wetlands. Wilderness Park is a haunt for birdwatchers, nature lovers and hikers.

The attraction of wildlife, however, is not always as idyllic as it appears.

"There is some thought that these constructed wetlands are pulling wildlife away from natural systems," says Bob Knight, an environmental scientist for CH2M Hill, an environmental engineering firm in Gainesville, Florida. "There's worry that wildlife in the [artificial] wetlands may be consuming toxins."

Only one such case has been recorded, in a marsh built in the 1960s in Kesterton, California. It was constructed to receive agricultural drainage that would gradually evaporate. But the accumulated toxins caused disease and mutations in waterfowl.

The Kesterton disaster created consternation not only among environmentalists but also among engineers involved in the business of constructing artificial wetlands. "We all need to keep our eyes open," says Donald A. Hammer, manager of planning and wastewater management for the Tennessee Valley Authority in Knoxville, Tennessee.

A Clear Future

At least one person believes that large northern cities will someday be able to use natural waste treatment systems, and he's constructed one as an experiment.

In a 30- by 120-foot greenhouse nestled in a grimy industrial wasteland a few miles from downtown Providence, Rhode Island, 16,000 gallons of raw sewage daily—the approximate output of 120 households—are rendered into water that is theoretically potable but has not been approved for drinking by the state. Inside the greenhouse, rows of huge translucent cylinders contain water hyacinths, watercress, bald cypress seedlings, ginger and philodendrons, as well as snails and tilapia (a fish species), which break down the waste as it flows from tank to tank.

This unique system is the brainchild of John Todd, Ph.D., a visionary and former scientist at the Woods Hole Oceanographic Institution. Such small facilities, though viewed dubiously by state regulatory boards, have the potential to create permanent oases in urban neighborhoods, to treat sewage cheaply and effectively and to recycle water.

The obstacles faced by any emerging technology result from human caution as often as ecological unknowns. But detractors serve an important purpose: They force the proponents of natural systems to make sure the alternative solutions to the nation's sewage problems are as safe and effective as possible.

All the News That's Fit to Eat

By Billy Allstetter

Environmental activists and celebrities who think they're environmental activists press the public to recycle newspapers. And so we do, in large numbers. Maybe the message should be changed to "Build more recycling plants," because the supply of old newspapers outstrips the nation's ability to reprocess it into usable paper products. Larry Berger, however, may have a solution: Feed the newspapers to cattle and sheep.

Newsprint doesn't contain much in the way of vitamins or minerals, but it does have plenty of cellulose, long chains of energy-rich glucose molecules.

Newsprint doesn't contain much in the way of vitamins or minerals, but it does have plenty of cellulose, long chains of energy-rich glucose molecules. While ruminants, like cattle and sheep, can digest the cellulose found in most plants, they can't break down the more tightly bound cellulose fibers in newspaper.

Berger, a professor of animal nutrition at the University of Illinois, has developed a process that makes yesterday's news easier to stomach. He shreds the newspapers (using only those with soybean oil–based ink, known to be safe for human consumption), treats them with water and 2 percent hydrochloric acid, and then boils the mixture for up to two hours. The heat and acid break apart the cellulose fibers enough for bacteria inside ruminants' stomachs to finish the job.

Gobbling Up the News

In preliminary experiments, Berger replaced about 20 to 40 percent of sheep's regular alfalfa feed with the newsprint/acid mix. The animals gobbled it up.

Americans now recycle about six million tons of newspapers, Berger says. About 30 million cattle live in the United States. "If they were fed a diet of 20 percent newspapers, they could easily consume all the newspapers that are recycled today."

Black and White to Squirm All Over

By John J. Fried

Nothing visible to the eye recommends Penn-Pro Cellulose Manufacturing Company, of Washington, Pennsylvania, as a company that is pushing the edge of the environmental-technology envelope.

It sits in a marginal industrial and commercial area marked by gravel-covered streets, old plants and empty parking lots.

Its manufacturing plant is a one-story, 400-by-500-foot building of rusting corrugated iron. Assorted mechanical castoffs constitute the landscaping.

Its machinery has a rescued-from-the-junkyard look about it. It is staffed by workers dressed in old clothes—no surprise given the dust and dirt that line the premises.

Its office is a beat-up old trailer out front. The furniture has Salvation Army written all over it. The premises smell of cat, in keeping with the black and white mouser roaming back and forth between a food dish and other destinations.

But in a nation awash in wastepaper but bereft of ideas for recycling it into useful and profitable products, PennPro, located about 30 miles south of Pittsburgh, stands as a company successfully finding one new use after another for the bundles of newsprint people dump on its doorstep every day and the truckloads of

misprinted papers a local publisher ships over regularly.

Other recyclers are content to recycle newsprint by having it deinked and then selling it back to publishers, says Reed McManigle of the Ben Franklin Technology Center of Western Pennsylvania, a state office that helps finance businesses with promising technology. "They just hope for more deinking plants to come along so they can sell more paper back for printing," McManigle says. PennPro "is the only one in this area making something new."

Take, for example, worm bedding.

Who would have thought that chewed-up newsprint would be to a worm what a good mattress is to a human being? PennPro's owner and chief (and only) research and development specialist, Dennis Kopcha, that's who.

Exporting a Waste

It wasn't a humane concern for the welfare of worms that led Kopcha, a mechanical engineer by training, to his discovery, but rather the understanding that keeping alive the wriggling critters that will lure fish to hook is a matter of great importance to a fishing enthusiast.

"It's hard for someone who doesn't fish to understand," Kopcha says with some disdain in his voice when explaining the demand for his product to someone who is a stranger to the sport, "but fisherman pay a lot of money, $2 a dozen, for night crawlers. If you are not careful, you could lose half of them." That fearsome death rate, he says, has not been mitigated by other types of worm bedding—including peanuts, peat moss, even dirt—that have come along.

> *Who would have thought that chewed-up newsprint would be to a worm what a good mattress is to a human being?*

And it was only extensive research, the 42-year-old Kopcha adds, that convinced him that newspapers, ground up just right and mixed with just the right additives—which he will not discuss—will make a happy home for worms waiting to take the Big Dip.

So happy, Kopcha says, that losses of worms kept in his bedding run to less than 1 percent. And so happy that sales of the two-pound bags containing the bedding added up to about 200 tons last year. Those sales, Kopcha adds, make him the second-largest manufacturer of worm bedding in the United States.

"People say, 'How do you do this?'" says Kopcha, who has begun to export his worm bedding to Germany. "Well, you devote five years of your life to it. You sweat blood. I tried different papers. Different grinds of paper. Different amounts of moisture. Then you have to develop the process. Then you have to market."

Kopcha pauses a moment in the recitation of the research and development process. "Lee Iacocca gets on TV and says that he developed a car in two years," he says, wonder clear in his voice. "I think that's amazing."

The worm bedding, of course, is not the only product that has enabled Kopcha to schedule the creation of a second assembly line of machinery dedicated to chewing up newspapers and reconstituting them into something else.

Among other things he has developed:

➤ A new form of insulation that can be sprayed into walls and ceilings.
➤ A mulch that can be used with fertilizer and grass seeds to green everything from highway dividers to landfills.
➤ A substitute for asbestos—which is now on the way out, largely as the result of Environmental Protection Agency (EPA) decree—used as a stabilizer in a wide range of products.

Kopcha's newspaper-based product will be used as an additive to give roof tar the body it needs to stay put and, as Kopcha puts it, "not run down the sides of your house." Two years ago, there was no such item in Pennsylvania, Kopcha says. "Now, we are running four to five days a week making tons of it."

WHAT YOU CAN DO

It takes one 15- to 20-year-old tree to make 700 grocery bags. Buy and use your own durable canvas or string bags—they're light, convenient to carry and can be used hundreds of times.

Earth Day 2000

Kopcha's father bought PennPro, a manufacturer of insulation, out of bankruptcy in 1985 because he was bored with retirement. Dennis Kopcha stepped in two years later. Today, the young Kopcha operates as the essential entrepreneur/inventor: a commercial loner, secretive, suspicious of outsiders, no more willing to talk about revenue or income than to talk about the chemicals he pours into his products.

At the same time, he has known when the time has come to go outside the confines of his beat-up trailer-office for help. As a result, he has sought advice from the National Environmental Technology Applications Corporation (NETAC), a group formed by the EPA and the University of Pittsburgh to boost the commercialization of environmental technologies. He also has touched the Ben Franklin Technology Center of Western Pennsylvania for about $200,000 in grants to develop his products.

The money has been forthcoming, says Robb Lenhart, director of program development for NETAC, because Kopcha's ideas are "detailed and convincing" and because "he has vision."

Recycling's Future

Kopcha's vision, though, does not blind him to the realities of the marketplace in which he is dealing. Recycling experts say the nation is generating 157.5 million tons of solid waste a year, of which approximately 11 percent is currently being salvaged for recycling.

By the year 2000, experts say, the amount of solid waste will rise to 192.5 million tons. Environmentalists hope that recycling by then will reclaim 25 percent. But Kopcha knows that even desire and imagination are not sufficient guarantee against the vagaries of recycling economics. All he has to do is look down the 400-foot length of his factory and take in the bales and bales of crushed plastics littering the ground.

A fellow entrepreneur, Kopcha says, was hoping to enter the plastic-recycling business. Then, the price of one grade of plastic the businessman was collecting dropped from 20 cents a pound to 4 cents, and the price of another grade of plastic dropped from 30 cents to 7 cents a pound. "He went right out of business just like that and left me to clean up his mess," Kopcha says.

Yet Kopcha retains his sense of humor and chuckles at the notion that he represents a ray of hope in the battle to find ways of recycling the estimated 87 million tons of wastepaper destined for landfills nationwide each year. Two years ago, Kopcha says, PennPro chewed up 600 tons of wastepaper. The amount has grown to 1,200 to 1,500 tons now and may soon reach 2,000 tons.

For all of his inventiveness and expanding product line, Kopcha maintains that PennPro is "only a pebble." The true definition of infinity, Kopcha laughs, "is how much [waste] paper is available."

Reprinted with permission from *The Philadelphia Inquirer*, November 18, 1992.

Supermarket Savvy

By Will Nixon

The sleepy village of Chappaqua in the suburbs north of New York City seems like an eternal 1950s. The white clapboard shops have no neon signs. The grassy traffic circle has home-made signs for high school events. The morning train unloads maids and babysitters from Harlem and Yonkers to the waiting station wagons, now driven by the "thirtysomething" generation of moms.

But the shelves of the local Gristede's supermarket are full of change. "Peanut butter is the perfect example," says Roberta Wiernik, leading a green shopping tour on behalf of the local League of Women Voters. "Four years ago, I only found it in glass jars—now it's only in plastic." The five women following her down the aisles have their own tales from the packaging revolution. Jeri Goldberg, visiting from the Larchmont chapter of the league, has given up on supermarkets for finding waxed paper sandwich bags. "I have to go to a restaurant supply store and buy a case of 1,000," she says, which gets shared with other mothers from her PTA.

A Nightmare of Overpackaging

The 1990s were touted as the decade when "green" would sweep across supermarkets the way "lite" did in the 1980s. The shelves

> ## WHAT YOU CAN DO
>
> Buy food in large sizes. Buying one large can of beans or corn uses less packaging than buying two or three smaller ones.

tell a different story. "Just when we're disappointed over Aunt Jemima's Pancake Express," Wiernik says, referring to new plastic bottles of already mixed pancake mix that save the trouble of stirring water and powder, "along comes Bisquick with their version. Give me a break." The booming convenience food market, as such items as the new pancake mix are known, coupled with the microwave revolution have created a nightmare of overpackaging.

But not everyone is buying it. Wiernick stops in front of the aseptic juice boxes. "It's very hard with kids, I know," she says. "But some mothers tell me Thermoses are making a comeback." And some better alternatives like Downy's [fabric softener] refill bottles are coming on the shelves. The Gristede's manager, who doesn't mind having his store used as a teaching lab, hasn't stopped carrying the products packed in plastics galore, but he offers enough variety that one can shop lightly on the planet with a little training.

The New Castle chapter of the league, which includes the village of Chappaqua, didn't invent green shopping tours, but they've done the most to promote them. Several hundred people have now followed the league guides down the aisles since the tours began in late 1990. Last spring [1991], Citibank donated money and a video crew to put the tour on a 20-minute tape that the New Castle chapter now distributes nationwide.

But when all is said and done, shopping is not the only answer. Wiernik concludes her hour-long tour by saying, "We should be green conservers and not just green consumers." (For more information, contact The League of Women Voters of New Castle, P.O. Box 364, Chappaqua, NY 10514.)

Reprinted with permission from *E—the Environmental Magazine*, January/February 1992. Subscriptions $20/year; P.O. Box 6667, Syracuse, NY 13217; (800) 825-0061.

EARTH CARE ACTION

Glass's Colorful Future

By Jeanne Trombly

Green bottles. Brown bottles. Clear bottles. Someday soon, this old troika of the supermarket shelves may give way to a dazzling rainbow of colored jars and bottles as a new technology comes into play. The bottles of tomorrow may be made of clear glass coated with a thin layer of colored acrylic gel.

To recyclers, who already have their hands full sorting three shades of bottles, this flood of colors sounds like a nightmare. But the new glass is designed to work in their favor, and may actually be a panacea for their current woes. The acrylic gel melts off during the recycling process, leaving clear glass behind. Good-bye to color sorting and the havoc of broken glass it often creates.

"At our Harlem facility, up to 60 percent of the glass ends up broken as unmarketable aggregate," says Steve Anderson of Resource Recovery Systems, which collects recyclables, "so we either landfill it or give it to the nearby glassphalt plant." Some glass recycling programs barely cover their costs because of breakage and sorting.

But glass problems are being translated into opportunity by Brandt Manufacturing Systems, the company pioneering the process of color-coating. Roberta Turner, Brandt's director of corporate development, says manufacturers can paint glass containers any color they want, from cherry red to sky blue, by dipping clear bottles in an acrylic coating that gets heat-cured to a hard finish.

At the recycling end, the glass goes into a furnace where the dye melts off, leaving the new glass ready to be recolored. Air emissions from the burned dyes are harmless, Brandt claims, pointing out that the same technique is already used with aluminum cans.

No More Landfilling

Some in the glass business wonder if the Brandt system can work for a glass manufacturing plant that handles 100 to 500 bottles a minute. Dr. Paul Graham of Owens-Brockway, the country's largest glass manufacturer, says, "We like the idea, but we don't think Brandt is ready to go commercial." Turner replies that Brandt is already negotiating with another large glass manufacturer. And after seeing the results of a state-funded feasibility study, Tom Kacandis, senior market development specialist at the New York State Office of Recycling, may soon challenge glass manufacturers to use color-coating. "Glass may be 100 percent recyclable, but we're not closing the loop when we're forced to landfill it," he says.

Besides the Brandt process, other approaches are being tested around the world. A vapor coating technique developed by a British glass manufacturer already has American consumers fooled. The popular Bombay Gin is actually packaged in clear glass with a sapphire-blue coating.

Reprinted with permission from *E—the Environmental Magazine*, January/February 1992. Subscriptions $20/year; P.O. Box 6667, Syracuse, NY 13217; (800) 825-0061.

A Man and His Landfill Recycler

By Verna Corriveau

How to cut landfill usage 25 percent by next year? That's the question keeping city managers awake nights from coast to coast. In the quest for a quick fix, most have embraced curbside recycling as the 60 percent solution—despite claims by numerous experts that a modest 5 percent to 10 percent reduction is more realistic.

According to inventor Dr. Clifford C. Holloway, however, consumers may soon be able to trash those three-bin, space-taking monstrosities invading kitchens and garages everywhere—without causing a stink about environmental irresponsibility. Holloway claims he has the cure to the country's municipal waste woes—an automated recycling system capable of reducing municipal waste volume by an estimated 80 percent to 90 percent.

The nucleus of the tri-patented Holloway Process is a large, rotating vessel that looks like the naked fuselage of an aircraft. Inside, steam and pressure "cook down" batches of organic garbage into cellulosic fibers. Cellulose packs a potential wealth of commercial uses—as potting soil and as a livestock food supplement, as well as in recycled paper and low-grade textiles. It also has a British thermal unit value of 7,000, roughly the equivalent of low-grade coal. As for the inorganics such as glass, metals and plastics, the system captures these through a combination of vibrating screens and ingenious classifying methods.

To cap it off, the Holloway Process is the environmentalists' fantasy—a system that produces virtually no effluent or leachate. A Welsh colleague once described the aroma of the cooked garbage as reminiscent of Yorkshire pudding.

Negotiations are currently under way to construct a high-tech processor near the Louisiana state capital of Baton Rouge. "This will be light-years ahead of the prototype," declares Holloway. "For one thing, the vessel will be larger, and it'll process continuously instead of in batches. The loading and classifying processes will be totally automated."

> *The Holloway Process is the environmentalists' fantasy—a system that produces virtually no effluent or leachate. A Welsh colleague once described the aroma of the cooked garbage as reminiscent of Yorkshire pudding.*

The 400-ton daily capacity of the new processor will make it almost six times faster than the prototype—well able to service a typical landfill of about 200 acres. At an estimated $12 million, the cost of the new processor runs at about 40 percent of the cost of developing an average-sized landfill. Unlike a landfill, however, the processor's useful life is extendable indefinitely. (For more information contact Dr. Clifford C. Holloway, Lloyd Patterson International, P.O. Box 994, Ormond Beach, FL 32175; 904-672-9146.)

From *Buzzworm, the Environmental Journal,* January/February 1992. Reprinted by permission.

 EARTH CARE ACTION

Seahawks Above, Worms Below

The Kingdome stadium, home of the Seattle Mariners and SeaHawks, has some new players—a team of red compost worms that quietly munch their way through food wastes while thousands of fans cheer in the bleachers overhead.

In 1987, the Kingdome started a recycling program that last year [1990] collected 87,000 pounds of cardboard, glass, metals and mixed paper. In April of this year, three plywood bins, each with a population of about 1,500 red wriggler worms, were added to compost food wastes disposed of by Kingdome concessionaires. The program has been so successful that in August, West Works Recycling of Bothell, Washington, supplied three new plastic bins made of reclaimed 55-gallon plastic drums.

A maintenance department volunteer tends the worms, feeding them up to 50 pounds of food scraps per week. Careful records are kept to see how large a bite worms can take out of the Kingdome's food wastes. No problems with odors or flies have been reported, and the resulting compost is used on Kingdome flower beds.

Worm bins are only one of the products introduced by West Works Recycling that utilize high-quality plastic containers that would otherwise be discarded. Included are rain barrels, composters and a variety of planters, all marketed through garden nurseries, retail outlets and direct to consumers. Because they are made of reclaimed materials, retail prices are kept surprisingly low and public response has been very positive. (For more details, contact West Works Recycling, Inc., 10425 N.E. 185th Street, Bothell, WA 98011; 206-481-3392.)

From *In Business,* November 12, 1991. Reprinted by permission.

Fashionable Fertilizer Solves a Disposal Problem for Zoos

To zoo visitors, magnificent beasts like rhinos and elephants symbolize the romance and danger of the wilds. But to zoo caretakers, these oversized animals can seem less like glorious emblems than giant manure machines.

Until recently, the huge piles of animal waste generated each day at the nation's zoos were relegated with pinched noses to piles that ended up in city landfills. But in line with the rising tide of environmentalism, zoo keepers and entrepreneurs are turning this once troublesome waste into compost for eager buyers.

The (800) I-LUV-DOO line at Zoo Doo, a compost company in Memphis, takes orders for gift packs of the compost. And the Woodland Park Zoo in Seattle is sold out of manures through the summer.

"Gardeners really like it," said Pierce Ledbetter, the founder of Zoo Doo. He said the product is twice as rich in nitrogen as cattle or horse manures, but not so strong that it can burn a plant, the way chicken manures do. "And anyway," he said, "people just get a real kick out of using rhinoceros doo."

No Seeds or Hormones

Zoo Doo compost is made of the manures of plant-eating animals, mostly rhinoceros and elephant. The animal feeds at the zoo, unlike farm feeds, keep the manures free of both weed seeds and hormones.

Zoo Doo's product sells for $10 a 15-pound bag, enough for 750 square feet of garden. Zoo Doo is comparable in price to other exotic composts but about twice as expensive as cow manure products, Ledbetter said. Zoo Doo is allowed to compost or decompose for a minimum of six months, by which time it has a coffee-ground-like consistency and no detectable odor.

Gardeners who use Zoo Doo say it also keeps away pests like deer. While odorless to the gardener, to deer and raccoons Zoo Doo seems to retain the imposing smells of rhino and elephant. Gardeners in Utica, New York, have gone even further, rejecting mere herbivore manures in favor of Siberian tiger manure from the zoo to keep deer and rabbits away.

Disposal Fees Saved

While zoo officials seem slightly amused by the popularity of the manures, they are mostly relieved to be rid of a costly and noxious problem.

Gabe Silva, building and grounds manager at the San Diego Zoo, said: "Over the many years of having animals here, we were transporting all of the manure from our enclosures to the city landfill. We probably averaged two and a half to three tons of manure a day, and then you think this has been going on for 50 maybe 75 years. That's a lot of manure."

With landfill costs rising, zoos can spend thousands of dollars a year getting rid of the waste. Sue Maloney, supervisor of grounds and facilities at the Woodland Park Zoo, said that by selling rather than dumping manure, the zoo had saved more than $32,000 in landfill fees a year.

Even before composting companies began selling these exotic excrements, zoos had used them on their own grounds. "I've used elephant manure for a long time," said Rick Pudwell, the general foreman and horticulturist at the Memphis Zoo, the first in a growing list of zoos to collaborate with Zoo Doo.

Pudwell said the manures provide nutrients and help open and aerate the soil. Many zoos, like the Bronx Zoo, compost the manures and use them for their own grounds.

In addition to saving space at landfills, zoos say they hope they are promoting composting and environmental awareness in general.

Standard Yellow Pencil to Turn "Green"

By Henry Stern

A pencil-making company gave students and office workers something new to chew on in April of 1992. Faber-Castell chose Earth Week to debut a pencil made of recycled newspaper and cardboard fibers instead of wood.

Officials with the company, based in Parsippany, New Jersey, tout the "American Eco Writer" as the first major development in pencil composition since erasers were added at the turn of the century. "There is a growing demand for recycled products," said chairman of the board Chris Wiedenmayer. "We're trying to stay ahead of the curve."

About 15 percent of the one billion pencils sold each year by Faber-Castell will use the recycled product made at the company's Lewisburg, Tennessee, plant, he said. Some two billion pencils are sold each year in the United States. The "green" pencil could save 75 tons of waste a year.

In addition to consumer demand for environmentally conscious products, Wiedenmayer also said Faber-Castell will benefit because the price of wood will rise as it becomes more scarce. Wiedenmayer said Faber-Castell hopes to eventually make all its pencils from recycled material, which he said costs slightly more than wood.

The pencil will sell around the midlevel of the company's price range between 10 cents and 25 cents, Wiedenmayer said. Other than an "environmentally" colored green eraser instead of a red one, the "green" pencil looks and feels like a standard pencil. "The pencils are made on the same machines," product manager Brent Gulick said. "It's a little bit heavier, but to the consumer there is no perceived difference."

Saving Trees

The recycled newspaper and cardboard is mixed with water and then pressed into grooved slats. Then the mixture is sandwiched around the lead. "That's great," said Jan Gottlieb, a manager at the Environmental and Occupational Health Science Institute in Piscataway. "Why do you need to cut trees down to make a pencil?"

Jeff Francis, a spokesman for a Stamford, Connecticut—based solid waste education

---

WHAT YOU CAN DO

Don't use throwaway products. Avoid single- or limited-use goods such as plastic razors, cigarette lighters and disposable cameras.

group called Keep America Beautiful, said businesses are trying to be environmentally responsible. "Many corporations, particularly consumer products companies, are more aware of the need to be responsible to the environment," Francis said. The need to cut costs and respond to consumer demand also have prompted the "buy green" movement, Francis said.

Why do you need to cut trees down to make a pencil?

"Just knowing a company is recycling makes people more favorable to that company," said Ronald L. Mersky, an associate pro-

fessor of civil engineering at Widener University. Mersky, a recycling consultant, said some companies have misled consumers toward that end. For example, some detergent companies insert a needless middle layer of recycled material in their plastic bottles so they can say the product uses recycled resources.

He said the estimated 75 tons of waste the "green" pencil will save each year is less than a big city generates in one day, but the effort should not be minimized. "It's not to say it's not a productive contribution," Mersky said.

"We're not going to save the world," Wiedenmayer said. "It's a drop in the bucket, but everybody should do their little bit."

From the *Associated Press*, April 13, 1992. Reprinted by permission.

Forestry Professor Finds Alternative to Foam Peanuts in Packaging

By Gordon R. Johnson

Jim Rice's wood curls could send peanuts and popcorn packing.

Not as a snack, perhaps. But Rice believes they make a better packaging material than the foam peanuts or pesticide-laced popcorn often used to cushion items shipped in boxes.

"Unlike foam peanuts, wood curls are perfectly biodegradable and people don't have to throw them away. They can be used as mulch around plants or as pet litter," said Rice, a professor in the University of Georgia School of Forest Resources. In his research, Rice has invented a machine that makes wood shavings "that are very light, resilient and recyclable and perfect for packing," said Dr. Janice Kimpei, associate director for technology transfer for the University of Georgia Research Foundation, Inc., who helped Rice apply for a patent on the machine.

Rice came up with the idea of using tightly curled wood shavings as packing material when he was asked to help develop a machine for producing the small wood curls used in fragranced potpourri. Rice estimated that his machine could manufacture resilient curls from scrap wood and compete with foam manufacturers, who charge about 50 cents per cubic foot for foam peanuts.

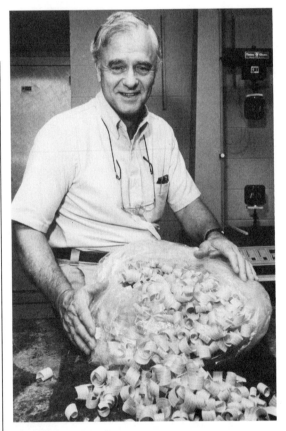

Jim Rice's wood curls are an inexpensive packing material that could replace foam peanuts and pesticide-laced popcorn. (Rick O'Quinn, University of Georgia)

WHAT YOU CAN DO

Buy loose fruits and vegetables.
Avoid shrink-wrapped produce in
Styrofoam trays. Try to do without
the plastic bags provided in supermarkets produce sections. If you
must use one, reuse it.

development as a new opportunity for companies that have sought more environmentally sound methods of packaging

In the past several years, some companies have experimented with new, biodegradable packing materials, especially popcorn, Kimple said. "Then bugs began to show up in the popcorn," she said. "Companies sprayed it with pesticide, but that presented a threat because people ate the popcorn. With wood curls, you have a packing material that is perfectly safe for people and the environment."

The curls can weigh as much as two-thirds more per cubic foot than the foam peanuts, but Rice said the difference would effect shipping costs very little. Kimple described the

From *Research Reports,* Spring 1992. Reprinted by permission.

This huge Los Angeles interchange symbolizes the automobile's prominent place in the American lifestyle. Unfortunately, motor vehicles account for almost one-quarter of the nation's emissions of carbon dioxide, the gas that threatens to warm the globe through a disastrous greenhouse effect. (Bruce Dale © 1983 National Geographic Society)

CLIMATE CHANGE BRINGS TROUBLE

By George F. Sanderson

We have come a long way in recent years toward realizing how extensively global warming will jeopardize both planetary and human health.

Carbon dioxide buildup—mainly from combustion of fossil fuels such as oil, gas and coal, and from clearing and burning of forests—is believed responsible for about half of this worldwide warming, while chlorofluorocarbons (CFCs), methane, ground-level ozone and nitrous oxide emissions account for the rest. Together, these gases act like glass in a greenhouse: They allow passage of incoming solar radiation but trap some of the outbound heat radiation from the earth.

One very significant aspect of global warming is the variety of effects it will have on human health—some will be subtle and indirect, others dramatic and direct. Many of the very factors contributing to global warming are themselves harmful to humans, such as the burning of fossil fuels: A typical automobile emits carbon monoxide, sulfur and nitrogen oxides, hydrocarbons, low-level ozone and lead, all of which are hazardous to health. According to a World Health Organization (WHO) task group on the potential health effects of climate change and the Intergovernmental Panel on Climate Change (IPCC), climate change will likely worsen air pollution—especially in heavily populated urban areas—by altering the composition, concentration and duration of chemical pollutants in the atmosphere.

One very significant aspect of global warming is the variety of effects it will have on human health—some will be subtle and indirect, others dramatic and direct.

Chlorine released by CFCs and bromine released by halons (used in fire extinguishers) both deplete stratospheric ozone. The use of CFCs and halons thus escalates the risk of skin cancer, eye cataracts, snow blindness and weakened immunity to a host of other illnesses by exposing humans to increased ultraviolet B radiation from the sun. "Skin cancer risks are expected to rise most among fair-skinned Caucasians in high-latitude zones," according to the IPCC. The WHO task group reached a similar conclusion, noting that "the incidence of nonmelanoma skin cancer could increase between 6 and 35 percent after the year 2050. The increase may be larger in the Southern Hemisphere, where total ozone depletions have been larger."

Health inside the Greenhouse

On the basis of present trends, scientists predict that greenhouse gases will warm the earth further by about 0.3°C (33°F) in each decade of the next century. This rise, faster than any experienced over the past 10,000 years, could increase the planet's mean temperature by 3°C (37°F) before the year 2100, making it warmer on average than it has been for 100,000 years.

This may not sound especially ominous, but left unchecked, global warming could alter rainfall patterns, flood vast areas of low-lying land as warmed seas rise (possibly as much as a meter) and drive countless species to extinction as fragile ecosystems collapse. A warmed planet will affect human health by disrupting food and fresh water supplies, displacing millions of people and altering disease patterns in dangerous and unpredictable ways.

The populations most vulnerable to the negative impacts of the greenhouse effect are in developing countries, in the lower-income groups, residents of coastal lowlands and islands, those living in semi-arid grasslands and those in the squatter settlements, slums and shantytowns of large cities.

Present strategies for immunization, coping with disease vectors or carriers, providing safe drinking water and improving nutrition are all based on existing climate regimes, ecosystems and solar-radiation and sea levels. These are all expected to change, but exactly how much cannot be predicted with any certainty, making it virtually impossible to adjust health and nutritional strategies now to take possible climate changes into account.

Humans adapt well to moderate changes in temperature and to occasional extremes. But this adaptive capacity—developed over many thousands of years—is relatively low in infants and the elderly; it rises through childhood and adolescence to reach a maximum that can be maintained up to about 30 years of age, then begins to decline.

When the Temperature Rises

Heat-related illness is one problem that will likely proliferate. Currently, the temperature in Washington, D.C., exceeds 38°C (100°F) on an average of 1 day per year; it rises above 32°C (90°F) about 35 days every year. "But by the middle of the next century, these figures could rise to 12 and 85 days respectively per year," according to the World Meteorological Organization. "The effect of such temperature rises on human health in Washington and similar cities throughout the world is difficult to predict. But there is no question that increased urban heat stress could come to claim many lives."

The same conclusion was reached by IPCC, which warned in June 1990 that the increase in deaths caused by a greater number of summer heat waves "would be likely to exceed the number of deaths avoided by reduced severe cold in winter."

As temperatures rise, the boundaries of the tropics may extend into the present subtropics, and parts of temperate areas may become subtropical.

A changing climate will also probably shift the range of conditions favoring certain pests and diseases, according to the final scientific statement issued by the second World Climate Conference in November 1990.

As temperatures rise, the boundaries of the tropics may extend into the present subtropics, and parts of temperate areas may become subtropical. This will allow the insects and animals that carry or cause many tropical diseases (e.g., mosquitoes, snails, etc.) to move poleward in both the Northern and Southern hemispheres. Some communicable illnesses, including those transmitted through air, water and food, could therefore become common in regions that once rarely knew them, with a possible rise in death rates.

Diseases such as malaria, hepatitis, meningitis, polio, yellow fever, dengue fever, tetanus, cholera and dysentery, which flourish in hot, humid weather, could increase, while those associated with dry, cold weather would be expected to diminish.

In a warmer climate, malarial mosquitoes and other disease carriers also may migrate vertically, up into formerly inhospitable highlands. This may be particularly hazardous in tropical highland areas where there is no natural resistance to malaria. Researchers in Kenya have already found malaria-carrying mosquitoes in areas where they were previously unknown.

Any changes in temperature, rainfall, humidity and storm patterns may affect insect- and animal-borne diseases in two ways. First, they may directly affect the carrier's range, longevity, reproduction rate, biting rate and the duration and frequency of human exposure. Second, they may modify agricultural systems or plant species, thus changing the relationship between carrier and host.

Development rates of mosquitoes, for example, would increase with warmer temperatures, provided these pests have wet areas in which to breed, and snail-borne diseases are likely to spread if global warming forces increased irrigation or causes

WHAT YOU CAN DO

Connect an automatic timer to the thermostat in your house. Set it to lower the temperature at times when the family is generally out. Timers are available from most heating contractors; some cost as little as $30.

people to migrate toward irrigation projects. Changed human migration patterns, along with increased temperature and rainfall, may extend the geographic range of hookworms, too.

Moreover, "warmer, humid conditions may enhance the growth of bacteria and molds and their toxic products, such as aflatoxins," cautioned the WHO task group. "This would probably result in increased amounts of contaminated and spoilt food."

Such changes would not be limited to developing countries. For example, in the United States, tick-borne diseases such as Rocky Mountain spotted fever and Lyme disease could spread northward. Americans could face the risk of five separate mosquito-borne diseases that have at present been virtually eradicated, according to Andrew Haines, a professor at University College and Middlesex School of Medicine in London.

Higher, Warmer Waters

As ocean temperatures rise and nutrients from agricultural fertilizers leach into rivers and coastal waters, toxic "red tides" may become more frequent, disrupting marine food stocks. This proliferation of minute marine organisms called dinoflagellates sets off a toxic chain reaction up the food chain: Incidences of food poisoning would increase when people eat tropical fish or shellfish that have eaten organisms that have eaten dinoflagellates.

Sea-level rise could spread infectious disease by flooding sewerage and sanitation systems in coastal cities, and increase the incidence of diarrhea in children. The flooding of hazardous waste dumps and sanitation systems could result in long-term contamination of croploads.

Rising, warmer seas may also disrupt marine habitats and aquatic food chains. Since fish constitute 40 percent of all animal protein consumed by the people of Asia, such a disruption of the marine ecosystem would affect the food supplies of many millions of people and dramatically increase protein deficiency and malnutrition.

Food shortages, reaching "famine proportions in some regions," could also follow the inundation of fertile coastal land by rising seas, the WHO task group noted. And the potential scarcity in some developing countries of food, cooking fuel and safe drinking water because of drought may further increase the extent of malnutrition, with "enormous consequences for human health and survival," according to the IPCC. The most serious implications are for Indonesia, Pakistan, Thailand, the Ganges Delta in Bangladesh and the Nile Delta in Egypt, all low-lying and densely populated.

Human Disruptions

Finally, changes in the availability of food and water as well as radical shifts in disease patterns could initiate large migrations of people. An increased number of "environmental refugees" would lead to overcrowding, social stress and instability, all of which may impair human health and increase health inequality between peoples of developed and developing countries.

Much more emphasis must be placed on research into how people contribute to and cope with climate change and on public awareness and education programs.

"Not only do we need more information about environmental conditions . . . we also need information about health conditions if we are to target our efforts and use our ever-limited resources to best serve health needs," notes Wilfried Kreisel, director of WHO's Division of Environmental Health. "Sad to say, environmental health globally suffers from informational malnutrition."

> *Day by day, the image of the world as the "global village" becomes more of a reality.*

Equally important in Kreisel's view is the global need to generate more and better human resources for environmental health, to develop more coherent environmental health policies and to influence not only the leaders of business and industry but also people in all walks of life to be more sensitive to health implications of their choices and decisions.

"Day by day, the image of the world as the 'global village' becomes more of a reality," Kreisel points out. And as all people are affected by environmental degradation, including that caused by global warming, "communication and sharing of resources among peoples is essential for the survival of the planet and our species."

From *The Futurist*, March/April 1992. Reprinted by permission.

 EARTH CARE ACTION

New Studies Predict Profits in Heading Off Warming

By William K. Stevens

Far from being ruinously expensive, efforts to head off a feared global warming could actually save money or even turn a profit in the long run, environmentalists, government analysts and even some business leaders and economists say.

The view is contrary to that of the Bush administration, which in international talks has resisted placing limits on emissions of carbon dioxide, a heat-trapping gas. The administration has found support from studies indicating that the costs of ultimately reducing emissions could run to tens of billions of dollars a year.

But now the opposite view is being pressed with increasing insistence. Citing studies of their own, advocates argue that reducing the emissions would force the economy to use energy more efficiently and at less cost in the long run, would free up large amounts of capital for expansion if the right governmental policies were followed and would make American business more competitive internationally.

"To ignore the economic opportunities is to fail to seize the moment, to become paralyzed by exclusive focus on one side of the economic ledger," Dr. Robert N. Stavins, an economist at the John F. Kennedy School of Government at Harvard University, said at a conference on the subject at the Smithsonian Institution in Washington.

A study by four environmental groups, coordinated by the Union of Concerned Scientists, concluded that aggressive action to lower carbon dioxide emissions by 70 percent over the next 40 years could cost the economy about $2.7 trillion. But it could also save consumers and industry $5 trillion in fuel and electricity bills, the study found, for a net saving of $2.3 trillion.

And unofficial studies by Environmental Protection Agency (EPA) analysts conclude that the gross national product would rise, not fall, over the next 20 years if emissions were reduced through a "carbon tax" on the extraction of coal, oil and natural gas—and the tax revenues were "recycled" back into the economy through investment tax credits for industry, thus spurring capital investment.

Competing studies consider different factors and operate on different assumptions, and

no consensus on the net economic effect of reducing carbon dioxide emissions has emerged. But the economic debate has been joined, and its outcome could be critical in the attempt to deal with a global environmental concern of the first magnitude.

*T*o ignore the economic opportunities is to fail to seize the moment, to become paralyzed by exclusive focus on one side of the economic ledger.

Carbon dioxide and other gases, chiefly methane, trap heat in the atmosphere much as glass panes trap it in a greenhouse. An international panel of scientists convened by the United Nations has predicted that if current emission rates continue, the earth's average surface temperature will rise by three to eight degrees Fahrenheit by the end of the next century. Many scientists believe this could have a catastrophic effect on climate.

The Energy Connection

Global warming and energy are inextricably linked because carbon dioxide, the most abundant and important greenhouse gas emitted by industrial society, is produced by burning fossil fuels like coal and oil. The United States, which lags behind other industrialized countries like Japan and Germany in energy efficiency, is the world's leading emitter of carbon dioxide. Thus, greater efficiency of energy use in this country is viewed as essential in trying to stabilize or reduce carbon dioxide emissions.

Analysis by the EPA suggests that modest energy-efficiency measures already under way will allow American per-capita emissions of carbon dioxide to stabilize at 1990 levels by the year 2000, at little or no cost to the economy. The European Community has called for overall emissions to be stabilized at 1990 levels by 2000; although the United States has opposed such a target and timetable, the EPA's analysis suggests that as a practical matter, it will be close to compliance anyway.

Sharp Reductions Urged

Many scientists believe that sharp reductions in carbon dioxide emissions—not just a stabilization—will ultimately be necessary. They say that at 1990 rates of emission, atmospheric concentrations of carbon dioxide would still grow. And in this context, the debate on the economic effects of reducing emissions takes on even greater importance. Answers will be needed if it becomes clearer in the next few years that the threat of warming is indeed serious.

One major study made public last December [1991], commissioned by Congress and carried out by the Department of Energy, concluded that reducing carbon dioxide emissions by 20 percent from 1990 levels would cost $95 billion a year in 2000. In general, its conclusions are consistent with those of a number of other government and private studies.

But the study is based on the assumption that the economy is already using energy in the most efficient way possible, said Alden Meyer, an energy analyst with the Union of Concerned Scientists. "That assumption is flat-out false," Meyer said.

He said the market contained many barriers to the adoption of more efficient practices and technologies. For instance, a landlord may not have an incentive to install more energy-efficient appliances or windows because tenants pay electricity bills. If the landlord was offered a subsidy to do so, the barrier could be overcome.

Incentives for Efficiency

Electric power companies in a few states are indeed paying customers to install more efficient lighting, cooling and heating systems, for example, in lieu of building new emission-producing power plants. The companies are allowed to recover the money by increasing rates. The saving to the economy comes in the form of lower electric bills for the building owner. The restraint on emissions comes both from lower energy use and not having to build a new power plant.

The study coordinated by Meyer found that if an aggressive policy to expand energy efficiency and switch to clean, renewable energy sources were pursued, the economy would reap a net saving of $2.3 trillion over the next 40 years and carbon dioxide emissions would be reduced by 70 percent.

Participating in the study along with the Union of Concerned Scientists were the Alliance to Save Energy, the American Council for an Energy-Efficient Economy and the Natural Resources Defense Council, all nonprofit organizations that advocate greater efficiency of energy use.

But an analyst at the Department of Energy, who spoke on condition of anonymity, said that while the department favored the removal of barriers to greater efficiency, consumer behavior was more complicated than the environmentalists' study assumes. That could make the barriers more difficult to remove, he said.

He said the environmentalists' study assumed that the economy would continue to shift away from manufacturing as it has in the past, which he called a questionable assumption. He said the study also assumed a lower basic growth rate for the economy: 2.5 percent a year as compared with 3 percent in the Department of Energy study. A higher growth rate means emission reductions are more difficult and costly to achieve.

Meyer criticized the administration's projected 3 percent growth rate as unrealistically high, a view that is shared by many private economists.

Analysts for the EPA say the Energy Department study does not reflect the ways in which, according to the unofficial study by their agency, the proceeds of a carbon tax could be recycled through the tax system to spur economic growth. The analysts found that if the tax revenues were used to reduce personal income taxes or the federal budget deficit, this would barely offset, if at all, the negative effect of the tax.

That negative impact would come when the coal and oil industries passed on the carbon tax to their customers in the form of higher

WHAT YOU CAN DO

The proper use of shades, draperies, curtains or blinds can reduce the heat from sunlight coming into your home by more than 75 percent in the summer.

Earth Day 2000

Relatively cheap energy in the United States has made this country less efficient and therefore less competitive worldwide.

prices. But if the revenue was used to encourage business investment through tax credits, they found, the effect of the carbon tax would be more than offset, producing a net gain for the national economy.

A combination of investment tax credits and personal income tax reductions—to make the tax more acceptable to consumers—could achieve substantial reductions in carbon emissions at no cost, according to the analysts.

Market-based solutions of this sort could also make the price of American goods and services more competitive with those of the more energy-efficient economies of Europe and Japan, said Dr. Stavins.

While the current antitax mood of the United States discourages the adoption of a carbon tax, at least for now, the European Community is seriously considering it as the centerpiece of its efforts to reduce emissions.

Relatively cheap energy in the United States has made this country less efficient and therefore less competitive worldwide, Stephan Schmidheiny, a Swiss industrialist, told the Smithsonian conference. Schmidheiny is the principal adviser for business and industry to Maurice F. Strong, the secretary-general of the U.N. environment meeting.

Shakeouts are painful, and the move to an energy-efficient market would undoubtedly produce some transitional pain, even if the end result is a more robust economy. Coal and oil producers and their employees, for example, could be heavily hit by a carbon tax, even if other sectors benefited from the recycling of the tax.

The Ultimate Cost

Beyond all this, another question arises: What would be the cost of not taking action if serious global warming occurs?

Few economists have addressed it. One, Dr. William D. Nordhaus of Yale University, has published a "best guess" estimate that a doubling of atmospheric carbon dioxide would reduce this country's gross national product by one-fourth of 1 percent. But Dr. Nordhaus notes that many important factors are not included in economic accounts. These include "human health, biological diversity, amenity values of everyday life and leisure, and environmental quality," he wrote in a 1990 monograph.

It may be impossible to place monetary values on some of these big factors, even though they are clearly valuable. Biologists, for instance, fear that global warming will catastrophically speed a steady loss of biological species that is now occurring worldwide. They also believe that the world's forests harbor untold billions of dollars' worth of biological riches that could be exploited for pharmaceuticals and other products without harming the environment. But their worth has not been gauged.

The effort to do so is "low on the totem pole of what is of interest to our discipline," Allen Sinai, chief economist of the Boston Company, Inc., said at a debate on development and the environment held in Cambridge, Massachusetts, under the sponsorship of Earthwatch, a Boston-based scientific research and education organization.

Some scientists believe economists should refrain from even making the attempt until more is known about the biological world. The tropical forests, for example, are mostly unexplored and only a fraction of species with economic value have been identified. Pending more knowledge, Dr. E. O. Wilson, a conservation biologist at Harvard University, said at the Cambridge debate, economists "should avoid fruitless and dangerous exercises in cost-benefit analysis which is far beyond their reach and join in the ethic of saving every scrap of biological diversity possible."

Breezing into the Future

By Dick Thompson

A decade ago, windmills promised to be a clean, reliable source of power that could help wean America from its dependence on dirty fuels and foreign oil. The idea of harnessing an energy supply that was free as the breeze generated enough megawatts of excitement to light up an entire new industry. Spurred by generous government tax incentives, investors poured more than $2.5 billion into U.S. wind projects during the early 1980s.

But enthusiasm was not enough to propel the dream into reality. "Wind developed a reputation for not working, and it had the stigma of a tax scam," says Robert Thresher, the wind-program manager at the National Renewable Energy Laboratory in Golden, Colorado. Eventually the problems caused power companies to back away. And by 1985, when the tax credits expired, the remaining wind towers began looking more and more like monuments to a lost cause.

Now, however, there's new energy in the wind. Engineers have used advanced technology to make wind turbines that are far more efficient and cost-effective than those of yesteryear. Says J. Michael Davis, chief of renewable energy programs at the U.S. Department of Energy: "These machines are real and reliable." Today's models are capable of meeting 10 percent of America's energy demand, and within 30 years, newer versions could provide for a quarter of the nation's power needs. Such fig-

ures have reenergized the manufacturers of wind-power equipment and attracted the interest of foreign competitors. Utilities are conducting wind surveys and starting pilot projects. And a new breed of wildcatter is scurrying to buy up wind rights—licenses to erect what may be the oil wells of tomorrow.

An American Technology

For years, the wind industry's goal has been to produce power at rates similar to oil's: roughly a nickel for a kilowatt. Machines now operating in California can produce energy at seven cents per kilowatt. In areas of consistent high winds, the next generation, currently being deployed, will bring that cost down to five cents by 1995, and more advanced designs are likely to shave off another penny by the year 2000. While many locales do not have enough wind to use the technology, enhancements already in the works will expand by a factor of 20 the area of land that can generate wind power profitably, according to experts at the National Renewable Energy Lab.

Wind's success says something about a dicey political issue: Should government tamper with free enterprise to nurture a new technology? The answer for renewable energy sources is definitely yes. Had manufacturers

92

and utilities not received state and federal assistance early on, the future of wind power would now be controlled by either Japan or Europe; both have consistently funded wind research. Today American technology dominates the field.

In a sense, wind power has come full circle. In the early 1900s, most of the electricity on U.S. farms was provided by windmills. Those were replaced during the 1930s when the Rural Electrification Administration wired the countryside. But the oil embargoes and environmental concerns of the 1970s prodded politicians to encourage the investigation of alternative energy sources. States began requiring their utilities to spend between 1 and 2 percent of profits on research, and the federal government added its generous tax credits for investments in renewables.

> *H*ad manufacturers and utilities not received state and federal assistance early on, the future of wind power would now be controlled by either Japan or Europe.

Unfortunately, the credits were for investment, not performance. Consequently, many wind-power machines seemed to be designed on an accountant's calculator to capture more deductions than breezes. Some towers were planted in fields of feeble winds. Others broke down with frustrating regularity. But a few companies persisted, and California in particular became the nursery for advanced technology. The state's hot central valleys are linked to the cool ocean by a series of gorges and valleys along the coast that act like wind

tunnels. It was in these natural labs that engineers began testing new designs.

Learning from Mistakes

The failures of the 1980s showed the researchers that they knew almost nothing about building machines that could withstand and harness the turbulence of wind. Early models used blades of a type originally designed for helicopters. Since wind pressure could vary considerably from one end of the blade to the other, the rotor would wobble wildly and eventually break off. Sudden gusts of wind could overpower the machine and burn out its energy-converting turbine. Some engineers tried solving the problems by building heavier machines, but that simply made them more expensive.

After much trial and error, researchers modified the contours of the blades; some, for instance, are thicker in the middle in order to provide more stability. Engineers put electronic sensors atop the towers that could constantly monitor wind direction and turn the machine to correct for changes. The sensors do not respond to every fluctuation, but when a computer calculates a sustained 15-degree shift, it signals for a turn into the wind. The leading American manufacturer, U.S. Windpower of Livermore, California, has built machines with electronic components that act as a giant surge protector, keeping sudden bursts of energy produced by gusts from overpowering the turbine.

Researchers also found that less than ideal placement of a windmill can have a major impact: Missing 10 percent of the wind can reduce power by 30 percent. Moreover, the arrangement of turbines within a wind "farm" is important because the wake produced by one windmill affects those around it. Computers are being used to simulate varied terrain and calculate how to produce the most energy.

The advances are slowly changing the way utilities evaluate the technology. "We look at it as a real competitive option," says Carl Weinberg, director of research for San Francisco–based Pacific Gas and Electric. Outside California, however, wind power still carries the burden of past failures. Even though a government survey found that ten midwestern states could more than meet all their electrical power needs from wind, no major wind projects are planned in the region for 1992.

Incentives Are in the Wind

But growing public concern over pollution from burning fossil fuels will increase the pressure for renewable energy. Several states are starting to require utilities to factor the cost

of environmental damage into the cost of power production. In California, where the process of calculating environmental cost is just beginning, wind power may be assigned a price 15 percent lower than that for energy from traditional sources.

Seven different proposals are before Congress to provide incentives for the purchase of new wind-turbines. Surprisingly, the energy industry itself is divided on the value of such incentives. Turbine manufacturers believe that wind should prove itself competitive without further special assistance. But utilities would like a tax credit to make investment more attractive.

Additional technological advances now on the drawing board are likely to make wind power even more appealing. Engineers plan to boost the towers in some areas higher than they are at present so that the machines can escape ground turbulence and tap more consistent winds. Lighter materials could reduce the cost of building the towers. And researchers are looking into ways to store excess energy produced during windy periods so that it could be banked for use on calmer days or during peak energy demand.

If wind power does not fulfill its promise as a major energy source by the end of the century, it will not be a failure of technology. It will be a failure of vision on the part of society to make the necessary commitment.

WHAT YOU CAN DO

Don't switch your air conditioner to a colder setting when you first turn it on. It won't cool the room any faster, and it will waste energy.

H₂ OH!

By Sam Flamsteed

On May 6, 1937, the German airship *Hinden-berg* glided onto the landing field at Lakehurst, New Jersey, burst into flames and crashed to the ground. The giant zeppelin had been filled with hydrogen, a very light gas that burns easily in air. All it took was a spark to end the age of the dirigibles.

Yet the most significant fact about the *Hindenberg* disaster, according to 41-year-old physicist Joan Ogden, was not that 35 passengers died, but that 62 survived. Most of the gas in the *Hindenberg*—millions of cubic feet—floated away before it could burn. People may think of hydrogen as a dangerous explosive, says Ogden, but in practical use the gas disperses so quickly that it's very difficult to maintain flammable concentrations.

Ogden repeats that point whenever the subject of hydrogen fuel comes up in conversation. And in her conversations, it comes up pretty often. Ogden, who works at the Center for Energy and Environmental Studies (CEES) at Princeton, believes hydrogen could and should be the nation's primary combustible and that the transition from conventional fossil fuels can begin as soon as the turn of the century. Working with CEES colleague Robert Williams, she's drawn up a plan for a hydrogen-based economy—not a vague, quixotic scheme, but an analysis that explores the generation, storage and transmission of hydrogen gas for use in cars and homes across the country.

No Pollution

In Ogden's scenario, vast arrays of solar cells set up in California, the Southeast and the deserts of the Southwest will generate hydrogen by electrolyzing water from underground aquifers. The gas will then be pumped into pipelines and distributed all over the country. Local sources of hydrogen, from, say, the gasification of biomass and electrolysis driven by wind power, will be added to the supply. Some of the hydrogen will be burned in power plants to generate electricity, some will be used directly for home heating and some will be used in cars and trucks, with hydrogen pumps replacing the gasoline pumps at service stations.

"Environmentally, hydrogen is nearly the ideal fuel," Ogden says. "Burning it releases zero carbon dioxide, zero sulfur, zero hydrocarbons, zero carbon monoxide, and zero particulates. All you generate is water vapor and a small amount of nitrogen oxides." Even the nitrogen oxides would be eliminated with the development of the fuel cell, a kind of battery that combines hydrogen and oxygen to produce an electric current.

Sounds great—so great that it's hard to believe no one's thought of it before. In fact, says Ogden, someone has. In the 1970s, when two major oil crises put a premium on alternative fuels, there was plenty of research into

95

hydrogen-burning vehicles. The problem, then and now, is that the most straightforward way to produce the gas is to pass an electric current through water, separating the H_2O into H_2 and O. Electricity, alas, is usually generated by burning coal or oil, which are both pollutants and nonrenewable, to boot.

"One obvious answer to the pollution question was to use photovoltaic cells, converting renewable sunlight into electricity," says Ogden. But at the time solar cells were too expensive and inefficient. So most researchers concluded that they would never be able to generate hydrogen economically enough to make it an important fuel.

By the early 1980s, however, solar energy experts had come up with more efficient technologies, including thinfilm solar cells that are easy and less expensive to mass produce. Williams decided in 1985, based on those advances, that it might be time to reexamine hydrogen power. At about the same time, he interviewed Joan Ogden, then a young physicist who was applying to the center under a one-year visiting professorship. Williams suggested she work on the hydrogen study if she won the grant—and she did.

Seeking Cost-Effectiveness

Ogden had been interested in energy issues since graduate school at the University of Maryland, where she did a thesis on the theory of how hot gases should behave in fusion reactors. After graduate school, she went to work at the nation's premier fusion lab: the Plasma Physics Laboratory at Princeton. "I was

trying to reconcile my theory with the experiments," she says, "and it was fun working with people who were taking real data."

But she also realized that the technical barriers to fusion were enormous, and she wanted to work on something whose practical applications were merely years away rather than many decades. "I wanted more control over what I worked on and over my life in general," she says. After two years she left to become a consultant, or, as she puts it, a freelance physicist. One of her new clients was CEES.

Environmentally, hydrogen is nearly the ideal fuel. Burning it releases zero carbon dioxide, zero sulfur, zero hydrocarbons, zero carbon monoxide and zero particulates.

"I found out that a lot of energy researchers there had a similar background to mine—they were trained in physics and wanted to address problems that would have broad implications for society," she says. She felt as if she had found her professional home, and eventually she joined the center's research staff.

In preparing the energy study, Ogden and Williams spent time not just at their computer keyboards, analyzing the performance of energy systems, but also out in the field, visiting laboratories involved in leading-edge solar-cell research and factories building more conventional cells for commercial use.

WHAT YOU CAN DO

Get a home energy audit to identify important measures for reducing energy use. An audit will determine if your home is poorly insulated and if the furnace is running improperly. Call your local utility company to find out how to get one.

Earth News

"The best experimental cells convert about 30 percent of the sunlight falling on them to electricity," she says, "but at present they're much too expensive for practical use. It's probably going to be more cost-effective to use thinfilm cells, which are cheap to manufacture. We estimate, based on a reasonable rate of technical progress, that they could be 12 to 18 percent efficient by the year 2000." The Department of Energy is even more optimistic, predicting efficiencies of 15 to 20 percent by then.

The Big Picture

Even with those improvements in efficiency, hydrogen's basic cost turns out to be higher than that of conventional fuels. But Ogden and Williams argue that there are other factors to consider. For example, unlike hydrogen, fossil fuels are subject to a number of indirect costs, including the expense of cleaning up pollution and treating pollution-related disease, and the security implications of depending on unstable Middle Eastern sources for oil imports.

Moreover, Ogden points out, hydrogen could make solar energy more practical as well. Solar cells can generate power only while the sun shines, and many population centers are far from the areas with the most intense sunlight. Somehow, the energy has to be stored for use at night and transmitted to where it's needed. The equipment for storing and transmitting hydrogen is much cheaper than batteries and power lines. "We've calculated that it is about one-fourth as expensive to pipe hydrogen across long distances as it is to transmit electricity the same distance," says Ogden.

She admits there are some problems with this scheme. One is that hydrogen has a lower energy content than gasoline: In fact, a car can go some 3,000 times farther on a gallon of gasoline than it can on a gallon of hydrogen. But you could improve the total mileage by pressurizing the fuel tank—and the thick walls of a pressure tank would have the added advantage of being extremely hard to rupture, even in a crash. Alternatively, the hydrogen could be stored in metal hydrides, compounds that absorb prodigious amounts of hydrogen and then release them when warmed.

Furthermore, Ogden and Williams calculated that to supply all the country's energy needs with hydrogen, an area one-fourth the size of New Mexico would have to be blanketed with solar cells. "That isn't exactly environmentally harmless," she says. But neither are the currently available alternatives, and Ogden thinks new emissions standards, prompted by a growing awareness of pollution's hazards, will make that apparent. The California Air Resources Board has already mandated a move to reduced-pollution vehicles, calling for 10 percent of all new cars to be totally emission-free by the year 2003. In that kind of regulatory climate, hydrogen starts to look pretty attractive.

Anticipating this growing interest, Ogden and Williams have gone on to the next stage of their analysis. "We've shown that hydrogen can be competitive," she says. "Now we're looking in a lot more detail at how you'd go about making it, and whether there are any problems we haven't thought of." While she occasionally works on other problems—one recent study analyzed the economics of burning sugarcane waste for fuel in Third World countries—Ogden's clearly saving her energy for hydrogen's cause.

The Cars of Tomorrow

By Sharon Begley

Was this what Henry Ford fomented a revolution for? So Washington bureaucrats could wrest from taxpaying Americans the right to big-finned, wide-axled chrome beauties? Foisting on citizens of the democracy tin cans with all the zip of golf carts? Not if the Big Three automakers had anything to say about it: In testimony against the proposed Senate bill mandating fuel efficiency, Ford predicted the measure would require "either all sub-Pinto–sized vehicles or some mix of vehicles ranging from a sub-sub-sub compact to perhaps a Maverick." Chrysler warned the bill "would outlaw . . . most full-sized sedans and station wagons." General Motors spun an equally bleak scenario.

The year was 1974, and the Big Three were responding to what would become the nation's first fuel efficiency standards for cars, established in 1975. Though the automakers didn't succeed then in halting the bill, their tune hasn't changed, and over the years it has helped curb further efforts at government control. "The automobile companies are trotting out the same scare tactics" against current pending legislation (the Motor Vehicle Fuel Efficiency Act) that they did 17 years ago, says National Wildlife Federation President Jay D. Hair. Efficient cars, they maintain, must be small and dangerous.

That notion is still the most widespread fallacy clouding the outlook for vehicles that are cleaner and less gasoholic. But it is hardly the only misinformation. Conventional wisdom also holds that cars of tomorrow exist only in blueprints. In fact, they already exist, and range from GM's sporty electric Impact to hundreds of United Parcel Service vans running on compressed natural gas to Volvo's prototype 100-miles-per-gallon (mpg) sedan. So why do these sleek marvels of efficiency seldom make it to the showroom?

Operation Desert Storm brought into stark relief just how much Americans will pay in lives and dollars to defend access to foreign oil. Most of that oil goes not to warm homes or power industry, but to move us around in 3,000-pound steel boxes.

That question assumed added importance in the wake of the war against Iraq. Operation Desert Storm brought into stark relief just how much Americans will pay in lives and dollars to defend access to foreign oil. Most of that oil goes not to warm homes or power industry, but to move us around in 3,000-pound steel boxes. Motor vehicles account for more than *half* of the nation's oil consumption, placing them "at the root of U.S. oil dependence," as

99

General Motors says that its Impact, a prototype slated for production, shows that electric cars may not be so far off after all. (Courtesy of General Motors Corporation)

Christopher Flavin of Worldwatch Institute puts it.

Greenhouse Fumes

National security is only the latest spur to the development of cars that use less or no gasoline. Thanks largely to cars and trucks, more than half of all Americans live in areas that do not meet clean air standards. "Motor vehicles generate more air pollution than any other single human activity," says Deborah Bleviss of the Washington-based International Institute for Energy Conservation. Vehicles also account for almost one-quarter of our emissions of carbon dioxide, the gas that threatens to warm the globe through a disastrous greenhouse effect. Without a revolution in how we drive, the pollution will get worse, for the car population is growing even faster than the human one.

To be sure, there has been some progress toward reducing cars' thirst for refined crude. Since 1976, fuel efficiency of new cars has doubled, now saving the country 2.5 million barrels of gasoline (and $110 million) every day, calculates Marc Ledbetter of the American Council for an Energy Efficient Economy. Moreover, estimates Ledbetter, "while a 40-mpg car would cost $600 more, it would save $2,000 over its typical life span." But fuel economy in new cars fell 4 percent between 1988 and 1990, as low gasoline prices lured buyers to bigger, more powerful gas guzzlers and the government did not tighten fuel-efficiency standards.

The reasons have nothing to do with technology—and everything to do with the marketplace. The U.S. Department of Transportation concluded in 1980 that technology *existing then* could produce a safe car that got 43 mpg. But, automakers point out, car buyers are not clamoring for high-mpg models. Such arguments infuriate conservationists, who blame automakers, not consumers, for resisting "every improvement in auto efficiency, safety and emissions," says Glenn Sugameli, an attorney with the National Wildlife Federation, "while spending a billion dollars a year to create, not respond to, consumer desires for overpowered muscle cars." Still, consumers can't help but notice that because gas is so cheap, a 35-mpg car saves an underwhelming $50 or so per year in fuel costs over a car that gets 30 mpg.

With little financial inducement to care about mileage, consumers and producers have shifted attention toward "high-performance engines with fast acceleration capabilities," says Bleviss. Traditionally, they have cared more about purchase price, styling, handling and zippiness—all qualities that Americans have tended to regard as their constitutional rights.

Well-Kept Secrets

Those attitudes may be changing. A recent survey conducted by the Union of Concerned Scientists found that 83 percent of the Americans polled favored an increase in federal automobile fuel-efficiency standards. However, automakers are reluctant to push new fuel-efficient technologies—and they staunchly oppose government efforts to mandate high-mpg cars. Such models, they claim, just sit on the showroom floor.

Yet automakers from Ford to Subaru have prototypes that get 60 to 100 mpg—though not always in cars destined for consumers. "It would be a mistake to think that just because a prototype is built it can be mass produced," says auto analyst Christopher Cedergren of J. D. Powers and Associates. "It's all part of the research and development game."

Automakers from Ford to Subaru have prototypes that get 60 to 100 mpg.

Among some of the latest gas-saving tricks (fuel savings estimates are based on research by the American Council for an Energy Efficient Economy):

➤ Doubling the number of valves per cylinder from the standard two, as do the Buick Regal, Honda Accord and Toyota Camry. Since these engines produce more horsepower than two-valve versions, fewer cylinders produce the same amount of power. Fuel savings: 10 percent.

➤ Using a continuously variable transmission, as on the Subaru Justy, which better matches engine speed to gear ratio. Estimated fuel savings: 4.7 percent.

➤ Reducing weight by about 10 percent (300 pounds), by substituting stronger light aluminum and fiber-reinforced plastics for heavy steel. Fuel savings: 6.6 percent.

➤ Replacing the pushrod engine with an overhead-cam engine that yields more power for less fuel. Fuel savings: 6 percent.

➤ Reducing drag with aerodynamic design. Fuel savings: about 4.6 percent.

➤ Turning off the engine when the car's not moving, as on the Volkswagen Eco-Golf diesel prototype. It gets 59 mpg in city driving rather than the 34 mpg for the regular Golf diesel; VW could begin limited commercial production next year.

➤ Installing intake valve control, which also cuts emissions. Honda and Nissan offer rudimentary systems; others are under development. Fuel savings: 6 percent.

These are among the strategies, according to Nevada Sen. Richard Bryan, that justify raising the CAFE (Corporate Average Fuel Economy) standards first introduced in 1975. A bipartisan bill Bryan introduced this year with Washington Sen. Slade Gorton would raise the standard for new cars to about 40 mpg by model year 2001 from the current 27.5. The payoff: our oil thirst would be reduced by 2.5 million barrels a day by 2005. "Current technologies can improve the gas mileage of the entire fleet," says Sugameli, "so Americans can choose from a full range of cars and light trucks that save money on gasoline and reduce pollution."

The other option for government action is to raise fuel prices, of course—and in fact the federal tax on gasoline rose a nickel in 1990.

But making people really care about fuel economy would require pricing similar to that of Europeans, with taxes high enough to push the price of a gallon of gasoline to $5 in some countries. As long as the market in this country does little to promote better fuel economy, "the only means left to encourage [that] is through government policymaking," says Bleviss.

Take the example of the Volvo LCP 2000. This prototype is more crashworthy than required by law, accelerates from 0 to 60 in 11 seconds and could be priced competitively. It gets 81 mpg on the highway and 63 mpg in the city. "It feels like a Honda Civic," says Lee Schipper of the Lawrence Berkeley Laboratory. "But it was allowed to die because fuel prices fell and Volvo felt that it wouldn't pay off. Most of the world today wants to buy cars that are bigger and more powerful."

That ambition is a strong indication of how far away this country is from the ultimate solution—breaking our love affair with the private car. Realistically, Americans aren't about to give up their vehicles; in 1988, 51 million more cars drove 450 billion more miles than in 1970. Even with tougher CAFE standards or incentives like rebates for fuel efficiency, a long-term solution may lie in vehicles that dispense with refined crude altogether. Among the alternatives to gasoline:

Compressed natural gas (CNG). The more than 500,000 CNG vehicles on the roads today (30,000 in the United States) show that CNG is a proven technology, one that reduces tailpipe emissions significantly and greenhouse gases somewhat. Depending on driving conditions, CNG vehicles spew 50 to 90 percent less carbon monoxide, 10 percent less carbon dioxide and 40 to 90 percent fewer reactive hydrocarbons (a principal cause of smog and ozone) than gasoline-powered vans, though emissions of nitrogen oxides are generally higher. Last February, GM announced a $40 million program with the Gas Research Institute to offer regular production CNG trucks by the mid-1990s.

Mass-produced CNG vehicles are expected to cost about $1,000 more than standard ones—mostly for the heavy, pressurized tanks needed to hold the CNG—but will save about $450 annually on lower fuel and maintenance costs. For about $2,500 to $3,500, a gasoline vehicle can be converted to run on CNG or to switch between gasoline and CNG with the flick of a dashboard switch. One drawback of today's CNG vehicles is that they have to be refueled every 200 miles or so—rather than 400 or so for gasoline.

Methanol, or wood alcohol. Thousands of methanol vehicles are on the road. "It's the alternative technology that's the furthest along and the most feasible," says Chrysler engineer D.C. Van Raaphorst. An alcohol fuel, methanol can be synthesized from wood, sugar, grains—anything containing carbon. Natural gas is the cheapest feedstock, and could produce fuel that is competitive with pre-Desert Shield gasoline prices.

Methanol can also be synthesized from coal; one idea for avoiding the emission of carbon dioxide from this process is to add hydrogen, turning CO_2 into yet more methanol. If the abundant high-sulfur coal in the United States were converted into methanol, some observers maintain, it could conceivably replace all of our Persian Gulf oil imports and more—though environmentalists point out that methanol produced from coal has a considerably higher impact on global warming than gasoline. What's more, new markets for high-sulfur coal encourage environmentally disastrous strip mining. And that, says Dave Alberswerth, National Wildlife Federation Director of Public Lands and Energy, who has long fought excessive strip mining, is "an *awful* idea."

Methanol's biggest advantage is that, as a liquid fuel, it does not require a major change in the distribution system or in car engines. Since it has a higher octane than gasoline (about 105), a methanol car is a muscle car—but a relatively clean one: It spews as little as 10 percent of the hydrocarbons that gasoline does. (Carbon monoxide and nitrogen oxide emissions are about the same as from gasoline.)

Most promising, methanol vehicles may not require carburetors, radiators or a cooling fan, argue Charles Gray and Jeffrey Alson of the Environmental Protection Agency, and could get by with a simplified fuel-injection system. Jettisoning these bulky components would allow for a sleeker, lighter car producing 60 to 80 percent less carbon dioxide than vehicles that burn gasoline, argue the two scientists in a report.

Methanol has drawbacks. It packs half the energy of gasoline, so unless drivers settle for reduced range, their cars will require a larger tank. Also, it is highly toxic, so precautions would be required at filling stations. Since it is so corrosive, it would require a fuel tank made of stainless steel or other resistant material, adding $300 to $500 to the sticker price. And it burns invisibly, which could lead to safety problems—though that could be remedied with a dye to color the flame.

By next year [1992], GM expects to deliver to the California Energy Commission, for use in its fleet, 2,000 Corsicas and Luminas that can run on either methanol or gasoline. Ford hopes to mass produce a "flex-fuel" Taurus sedan by 1993. Volkswagen will produce about 100 flex-fuel Jettas this year and another 200 the following year.

Electric. The California Air Resources Board has mandated that, by 1998, 2 percent of the approximately two million new vehicles sold in the state be electric; by 2003, the figure must be 10 percent. That requirement alone assures a market: Clean Air Transport will offer 30,000 electric vehicles in the state by 1995.

Already, buyers can purchase a custom-made electric car for little more than a standard gasoline car off the corner lot. Solar Electric of Rohnert Park, California, offers vehicles ranging from three-wheel electric mopeds ($995, speed up to 15 miles an hour) to converted Fiat X-19s ($7,000, 60 mph). Southern California Edison, working with Swedish and British firms, expects 10,000 electric four-passenger cars to hit the roads in California by 1995. General Motors has developed the sporty Impact, which travels 120 miles on a single charge.

One drawback is battery technology. "Current lead-acid batteries are more than adequate for our first generation of electric cars," says GM President Lloyd Reuss. "But because lead-acid batteries are heavy and short-lived, we'll need to develop advanced energy systems."

Better batteries are in the works. Chrysler's TEVan, for example, with a nickel-iron battery, is scheduled for production in the mid 1990s: It will get 120 miles on a charge and reach 65 miles an hour. Ford and BMW are both working on models powered with sodium-sulfur batteries. Tokyo Electric Power Company is developing a car powered by nickel-cadmium batteries. Detroit's Big Three pledged $35 million to a consortium for developing better batteries.

> *C*learly, the cars of the future can be ready to roll as soon as buyers want them. The challenge for lawmakers is to use the carrots and sticks of regulations and rebates, policy and penalty, to get them moving off the assembly line.

Even if batteries are improved, their use won't eliminate the burning of fossil fuels, as much of our electricity comes from coal. Though they would ease urban pollution, "nobody should kid themselves that this is a 'clean' technology," says National Wildlife Federation's Alberswerth. "We will simply be substituting stack gas emissions from coal-fired power plants for tailpipe emissions from cars."

Hydrogen. This technology is the furthest from commercialization, but, if produced by solar energy from water, it would make driving almost as benign as bicycling. Among the obstacles: hydrogen must be stored either at $-423°F$ to keep it liquid, or in metal hydride tanks in which the gas is stored in metal powders.

WHAT YOU CAN DO

Use your fan. Ceiling fans consume as much as 98 percent less energy than air conditioners.

Earth News

Clearly, the cars of the future can be ready to roll as soon as buyers want them. The challenge for lawmakers is to use the carrots and sticks of regulations and rebates, policy and penalty, to get them moving off the assembly line. More funding would help, too. In 1990, the U.S. government spent a total of $194 million on all energy-efficiency research. That came to less than the bill for one day of Desert Storm.

From *National Wildlife,* August/September 1991. Reprinted by permission.

A group of Australian scientists returns to their warm "apple" hut after trudging out into the subzero climate of the world's coldest continent. It was in Antarctica that scientists first discovered a hole in the earth's protective ozone shield. (Tom Stack & Associates © Dave Watts)

THE OZONE VANISHES

By Michael D. Lemonick

The world now knows that danger is shining through the sky. The evidence is overwhelming that the earth's stratospheric ozone layer—our shield against the sun's hazardous ultraviolet rays—is being eaten away by man-made chemicals far faster than any scientist had predicted. No longer is the threat just to our future; the threat is here and now. Ground zero is not just the South Pole anymore; ozone holes could soon open over heavily populated regions in the Northern Hemisphere as well as the Southern. This unprecedented assault on the planet's life-support system could have horrendous long-term effects on human health, animal life, the plants that support the food chain and just about every other strand that makes up the delicate web of nature. And it is too late to prevent the damage, which will worsen for years to come. The best the world can hope for is to stabilize ozone loss soon after the turn of the century.

If any doubters remain, their ranks dwindled in February of 1992. The National Aeronautics and Space Administration (NASA), along with scientists from several institutions, announced startling findings from atmospheric studies done by a modified spyplane and orbiting satellite. As the two craft crossed the northern skies in January, they discovered record-high concentrations of chlorine monoxide (ClO), a chemical by-product of the chlorofluorocarbons (CFCs) known to be the chief agents of ozone destruction.

> The evidence is overwhelming that the earth's stratospheric ozone layer—our shield against the sun's hazardous ultraviolet rays—is being eaten away by man-made chemicals far faster than any scientist had predicted.

Although the results were preliminary, they were so disturbing that NASA went public a month earlier than planned, well before the investigation could be completed. Previous studies had already shown that ozone levels have declined 4 to 8 percent over the Northern Hemisphere in the past decade. But the latest data imply that the ozone layer over some regions, including the northernmost parts of the United States, Canada, Europe and Russia, could be temporarily depleted in the late winter and early spring by as much as 40 percent. That would be almost as bad as the 50 percent ozone loss recorded over Antarctica. If a huge northern ozone hole does not in fact open up in 1992, it could easily do so a year or two later. Says Michael Kurylo, NASA's manager of upper-atmosphere research: "Everybody should be alarmed about this. It's far worse than we thought."

And not easy to fix because CFCs are ubiquitous in almost every society. They are used in refrigeration and air conditioning, as cleaning solvents in factories and as blowing agents to create certain kinds of plastic foam. In many countries CFCs are still spewed into the air as part of aerosol sprays.

A Call for Faster Action

Soon after the ozone hole over Antarctica was confirmed in 1985, many of the world's governments reached an unusually rapid consensus that action had to be taken. In 1987 they crafted the landmark Montreal Protocol, which called for a 50 percent reduction in CFC production by 1999. Three years later, as signs of ozone loss mounted, international delegates met again in London and agreed to a total phaseout of CFCs by the year 2000. That much time was considered necessary to give CFC manufacturers a chance to de-

velop substitute chemicals that do not wipe out ozone.

But the schedule now seems far too leisurely. NASA's grim news early in 1992 spurred new public warnings and calls for faster action. In Denmark an Environment Ministry spokesman went on television to urge fellow Danes not to panic—but to use hats and sunscreen. German Environment Minister Klaus Topfer called on other countries to match Germany's pledge to stop CFC production by 1995. Greenpeace activists in Britain met with Prime Minister John Major and implored him to halt the manufacture of all CFCs immediately.

The U.S. Congress passed a law in 1990 that called for an accelerated phaseout of CFCs if new scientific evidence revealed a greater threat to ozone than expected. In February 1992 the Senate, by a 96–0 vote, found the evidence alarming enough to justify a faster phaseout. "Now that there's the prospect of a hole over Kennebunkport," Sen. Albert Gore of Tennessee said, "perhaps Bush will comply with the law." William Reilly, administrator of the Environmental Protection Agency, said that the United States might seek to end CFC production as early as 1996.

The vital gas being destroyed is a form of oxygen in which the molecules have

---

WHAT YOU CAN DO

When your car air conditioner needs repair, go to a garage that recycles CFCs instead of releasing them into the atmosphere.

three atoms instead of the normal two. That simple structure enables ozone to absorb ultraviolet (UV) radiation—a process that is crucial to human health. UV rays can make the lens of the eye cloud up with cataracts, which bring on blindness if untreated. The radiation can cause mutations in DNA, leading to skin cancers, including the often-deadly melanoma. Estimates released last week by the United Nations Environment Program predict a 26 percent rise in the incidence of nonmelanoma skin cancers worldwide if overall ozone levels drop 10 percent.

Excess UV radiation may also affect the body's general ability to fight off disease. Says immunologist Margaret Kripke of the M. D. Anderson Cancer Center in Houston: "We already know that ultraviolet light can impair immunity to infectious diseases in animals. We know that there are immunological effects in humans, though we don't yet know their significance."

Just as worrisome is the threat to the world's food supply. High doses of UV radiation can reduce the yield of basic crops such as soybeans. UV-B, the most dangerous variety of ultraviolet, penetrates scores of meters below the surface of the ocean. There the radiation can kill phytoplankton (one-celled plants) and krill (tiny shrimplike animals), which are at the very bottom of the ocean food chain. Since these organisms, found in greatest concentrations in Antarctic waters, nourish larger fish, the ultimate consumers—humans—may face a maritime food shortage. Scientists believe the lower plants and animals can adapt to rising UV levels by developing cell pigments that absorb UV radiation. But that works only up to a point, and no one knows what that point is.

Twilight Soccer

The impact of ozone loss will be felt first in Antarctica, where levels of the gas have been severely depleted each spring for several years. Populations of marine organisms are not shrinking so far, but they have begun to produce UV-absorbing pigments. In Australia, scientists believe that crops of wheat, sorghum and peas have been affected, and health officials report a threefold rise in skin cancers. There are anecdotal reports of more cancer in Argentina, too. While no increase in cancers or cataracts has shown up yet in Chile or New Zealand, experts note that these diseases can take years to develop.

Many people are reducing their risks. In Punta Arenas, Chile's southernmost city, some parents keep their children indoors between 10 A.M. and 3 P.M., and soccer practice has been moved from midafternoon to later in the day. The Australian government issues alerts when especially high UV levels are expected, and public service campaigns warn of the dangers of sunbathing, much as U.S. ads counsel people not to smoke. In New Zealand, schoolchildren are urged to wear hats and eat their lunches in the shade of trees.

Solar Pressure Cooker

Scientists are also concerned about the potential effect of ozone depletion on the earth's climate systems. When stratospheric ozone intercepts UV light, heat is generated. That heat helps create stratospheric winds, the driving force behind weather patterns. Sherwood Rowland, a chemist at the University of California at Irvine, who first discovered the dangers of

CFCs, says, "If you change the amount of ozone or even just change its distribution, you can change the temperature structure of the stratosphere. You're playing there with the whole scheme of how weather is created."

Weather patterns have already begun to change over Antarctica. Each sunless winter, steady winds blow in a circular pattern over the ocean that surrounds the continent, trapping a huge air mass inside for months at a time. As the sun rises in the spring, this mass, known as a polar vortex, warms and breaks up. But the lack of ozone causes the stratosphere to warm more slowly, and the vortex acts as a sort of pressure cooker to intensify chlorine's assault on ozone molecules.

When Rowland and his colleague, Mario Molina, issued the first ozone alert back in 1974, they had no idea that depletion would be particularly severe in Antarctica or in any other part of the world. What they did predict was the CFCs would not disintegrate quickly in the lower regions of the atmosphere. Instead the hardy chemicals would rise into the stratosphere before dissociating to form ClO and other compounds. The highly reactive chlorine would then capture and break apart ozone molecules. Each atom of chlorine, it was later determined, could destroy up to 100,000 molecules of ozone—at a far faster rate than gas is replenished naturally.

But Rowland and Molina had deduced only the broadest outlines of the process. The details had to wait until the mid-1980s, when atmospheric scientists realized belatedly that while worldwide ozone levels had declined somewhat, there was an enormous deficit in Antarctica every year. Determined to understand whether CFCs were the culprit, NASA mounted a series of flights from Punta Arenas into the Antarctic in 1987. They revealed unusually high concentrations—up to one part per billion—of ClO. They had found the smoking gun Rowland and Molina had predicted.

Rowland and others figured it was a combination of factors that made the ozone over Antarctica particularly vulnerable. First, the polar vortex collects CFCs that waft in from the industrialized world. Second, the superfrigid air of the Antarctic night causes clouds of tiny ice crystals to form high up in the stratosphere. When the CFCs break down, the resulting chemicals cling to the crystals, where they can decompose further into ClO, among other substances. And finally, when the sun rises after the long winter night, its light triggers a wholesale demolition of ozone by chlorine monoxide.

Coming to America

In Antarctica winds circulate unimpeded over the frozen landmass. In the north, though, the polar vortex is less well defined. Winds travel alternately over land and water, whose differing temperatures disrupt the smooth flow of air. The vortex wobbles and sometimes breaks up entirely. Moreover, the Arctic stratosphere is not as cold as that over the Antarctic, and ice clouds are less likely to form. So while scientists knew that some ozone destruction should take place, they presumed it would not be nearly as severe as the southern hole. A reanalysis of ten years' worth of ground-based and satellite data, completed last year [1991], revealed a relatively mild but widespread depletion over the Northern

Hemisphere, with losses of 4 to 8 percent over much of the continental United States.

When NASA's ex-spy plane, the ER-2, began a series of flights out of Bangor, Maine, in October [1991], it quickly became clear that something strange was happening. For one thing, volcanic ash, lofted into the stratosphere from last year's Mount Pinatubo eruption, was evidently taking the place of ice crystals, giving CFC by-products the platform they needed for their chemical reactions. Moreover, the scientists found that naturally occurring nitrogen oxides, compounds that tend to interfere with and slow down these reactions, were virtually gone from the atmosphere. Why? Besides enhancing the reactions that create ozone-destroying forms of chlorine, explains Susan Solomon, a chemist with the National Oceanic and Atmospheric Administration, "the volcanic aerosols provide a surface for chemical reactions that suppress nitrogen oxides."

Another flight that took off from Maine in January [1992] provided the clincher. The polar vortex had temporarily dipped as far south as Bangor—"It was almost as if we were deployed over the North Pole," says geophysicist Darin Toohey of the University of California, Irvine—just in time for the sensitive instruments on board to detect ClO in a world-record concentration of 1.5 parts per billion. Data from the Upper Atmosphere Research Satellite had already found comparable levels of ClO over Northern Europe, and the evidence pointed to a potential ozone loss of 1 to 2 percent a day.

Even with all these factors in place, there is still one element necessary before a certified ozone hole can form: the sun. If the polar vortex breaks up before the sun rises after months of darkness to trigger the reaction, there will be no hole this year. If the vortex holds together until late February or early March, keeping its brew of dust particles and chemicals intact, ozone levels will almost certainly drop. Harvard chemist James Anderson says, "We are now protected only by the hope of a rapid breakup of this vortex." But even if the hole does not appear this spring, says Anderson, it will almost certainly appear within the next few years. (*Editor's Note:* Ozone depletion over North America wasn't as bad in the spring of 1992 as it could have been, although fairly significant depletion did occur over arctic regions of Canada.)

---

*L*ife in the far north could come to resemble that in Australia, with ozone alerts and stern warnings to wear sunglasses and sunscreen.

When it does, the area of greatest ozone depletion and greatest danger will most likely to be north of 50° north latitude, a line that nearly coincides with the U.S.–Canada border and also takes in all the British Isles, Scandinavia, the Netherlands and much of Belgium, Germany and Russia. Regions farther to the south could be affected too, albeit not so severely. Life in the far north could come to resemble that in Australia, with ozone alerts and stern warnings to wear sunglasses and sunscreen.

Some scientists are equally concerned about the smaller but worsening ozone loss at mid-latitudes. The mechanism behind polar ozone holes was not predicted before its discovery. Could there be an undiscovered reason for ozone to vanish over temperate zones as well? Maybe so. On one flight, the ER-2 swooped south instead of north. Says Anderson: "We discovered to our shock that there was ClO all the way down to the Caribbean." It was a very thin layer with concentrations of only 0.1 part per billion—but this was much higher than anyone had predicted.

Elusive Cure

No one is sure just how such concentrations of the chemical got there or whether it is destroying ozone. It may be that some of the ClO-rich air from the polar vortex has split off and headed south on its own—a phenomenon that has been observed in the past. And while ozone depletion has not been directly observed, the chemistry over the Caribbean appears to be right. There is ClO; there are plenty of dust particles from Pinatubo; there is sunlight. NASA's Kurylo thinks significant ozone loss is in fact happening in the tropics. Says Harvard's Anderson: "This is cause for extreme concern. It is the mechanism we most fear."

What also frightens scientists is the fact that CFCs remain in the atmosphere for decades after they are emitted. In their original research, Rowland and Molina estimated that CFCs can last 100 years or more. Even if CFC production stopped today, researchers believe that stratospheric levels of chlorine would continue to rise, peaking during the first decade of the next century and not returning to anything like natural levels for at a least a century.

The ozone story is a tragic saga of doubt and delay. Rowland recalls that for several months after his original ozone paper was published in 1974, "the reaction was zilch." It was not until 1978 that the United States, but not most other countries, banned the use of CFCs in hair sprays and other aerosols.

Not until the Antarctic ozone hole was confirmed in 1985 did nations get serious about curbing all uses of CFCs. By now as many as 20 million tons of these potent chemicals have been pumped into the atmosphere.

World leaders should remember ozone when they think about other threats to the planet. If they always wait until there is indisputable evidence that serious damage is occurring, it may be much too late to halt the damage. Consider the widespread scientific predictions of global warming from the greenhouse effect. No one knows for sure that anything terrible will happen. But humanity has boosted the amount of carbon dioxide in the atmosphere by at least 25 percent. It is reckless to subject nature to such giant experiments when the outcome is unknown and the possible consequences are too frightening to contemplate.

At least nations now seem to agree on a crash effort to save the ozone. But the cure will not be instantaneous. The world may not know for decades how costly the years of recklessness will be. And whether children should be afraid to look up.

As the ozone gets thinner, people—especially young children—may have to cover up year-round to guard against harmful radiation from the sun. (Sally Shenk Ullman)

 EARTH CARE ACTION

Hats On!

Ozone depletion is cause for caution, but it's no reason to stay barricaded indoors or put on an astronaut suit before venturing outside. Excessive exposure to the sun's ultraviolet (UV) rays has always been dangerous; the ozone problem just adds to the risk. Michael Kurylo of the National Aeronautics and Space Administra-

tion says, "We're not talking about a single exposure to a death ray. It takes repeated exposure over long periods of time."

Even if there were no atmospheric damage, an estimated one-sixth of all Americans would still develop skin cancer during their lifetime. Most cases are curable, if detected early. The 4 to 8 precent loss of ozone over the past decade could raise the risk at least 15 percent. A significant increase in cataracts, which now afflict one of every ten Americans, could also occur.

As the ozone depletion gets worse, health risks will rise, but the odds of getting cancer or cataracts can be dramatically reduced by following guidelines that doctors recommended long before ozone depletion became a big issue. Their suggestions:

➤ When out in the sun for prolonged periods, wear protective clothing. That means choosing fabrics that have a tight weave and donning a wide-brimmed hat. A baseball cap is not adequate because it leaves the delicate rims of the ears exposed.

➤ In summer, when comfort calls for shorts and T-shirts, use a broad-spectrum sunscreen with a sun protection factor (SPF) of at least 15.

➤ Minimize the time spent in the sun between 10 A.M. and 3 P.M.

➤ Wear sunglasses when outdoors in bright sunlight. Ask for glasses that are treated to absorb UV radiation or that meet the American National Standards Institute Guidelines for eyewear. Poorly designed sunglasses that do not block UV rays could do more harm than good. Under dark lenses, the pupils dilate, making it easier for UV light to damage the delicate membrane of the retina.

From *Time*, February 17, 1992. Copyright © 1992, The Time Inc. Magazine Company. Reprinted by permission.

 EARTH CARE ACTION

How Do You Patch a Hole in the Sky?

By Philip Elmer-Dewitt

Think for a moment about the world's one billion refrigerators and its hundreds of millions of air conditioners. Picture mountains of foam insulation, seat cushions, furniture stuffing and carpet padding. Imagine streams of cleaning fluids, rivers of industrial solvents, wafting clouds of aerosol spray.

Ridding the planet of the millions of tons of ozone-depleting chemicals contained in that vision is not just a big job; it may be the biggest job the nations of the world have ever taken on. In the 60 years since Du Pont began marketing the miracle refrigerant it called Freon, chlorofluorocarbons (CFCs) have worked their way deep into the machinery of what much of the world thinks of as modern life—air-conditioned homes and offices, refrigerated grocery stores, climate-controlled shopping malls, squeaky-clean computer chips. Extricating the planet from the chemical burden of that high-tech life-style—for both those who enjoy it and those who aspire to it—will require not just technical ingenuity but extraordinary diplomatic skill.

The technical challenge is relatively straightforward. The goal is to find substances and processes that can replace CFC-based systems without doing further harm to the stratosphere—an endeavor that is well under way. In fact, it may turn out to be easier than anyone expected. Except for medical aerosols, some fire-fighting equipment and certain metal-cleaning applications, there are now effective substitutes for virtually every ozone-depleting chemical. Some cost quite a bit more, and others pose different, if less severe, environmental problems. But in a surprising number of cases, the new processes are actually cheaper and better than the old.

> *E*xcept for medical aerosols, some fire-fighting equipment and certain metal-cleaning applications, there are now effective substitutes for virtually every ozone-depleting chemical.

Replacing CFCs in newly built equipment, however, is only half the job. Virtually every

---

WHAT YOU CAN DO

When buying solvents, sprays, photo- and electronic-equipment cleaners, look at the labels. Do not use products that contain ozone-depleting chemicals such as CFCs, carbon tetrachloride, methyl chloroform and halons.

California magazine

---

existing refrigerator and air conditioner is a CFC reservoir. The chemicals are not a problem as long as they continue to circulate within an appliance. But if the machine is carelessly drained, junked or damaged, the CFCs can escape to attack the ozone. The real task for those countries that invested heavily in CFCs in the past is to develop systems for recovering and recycling the chemicals they have already used.

The diplomatic challenge is trickier. For the United States, Europe and other industrialized regions to do right by the stratosphere is one thing. They bear direct responsibility for most of the damage that has been done, and they can best afford the costs attached to switching technologies. But what about the countries of the Second and Third Worlds? Many of them are just beginning to enjoy the comforts of CFC technology, and they cannot easily pay for a changeover.

The progress made so far is encouraging. According to the United Nations Environment Program, which oversees the Montreal Protocol, there has been a 40 percent drop in CFC consumption since 1986, largely because of accelerated phaseouts in industrialized countries. There has been a similar reduction in the halons—the ozone-hostile chemicals used in fire fighting. In 1990 the Montreal Protocol was broadened to include two potent industrial solvents not covered in the original agreement: methyl chloroform and carbon tetrachloride. U.N. officials are now convinced that the developed world will have stopped making the most prevalent kinds of ozone depleters by 1995 or 1997, depending on the particular substance, and that developing countries may be able to catch up in five to eight more years— not the ten extra years once anticipated.

Worldwide Initiatives

Some of the countries that resisted CFC controls at first are taking the lead today— sometimes to their own surprise. Germany, which was dragged by its heels to the initial Montreal meeting, became the first country to establish a system for recycling CFCs from discarded refrigerators. Sweden, Switzerland and the Netherlands are among other countries working on their own refrigerant-recycling programs. Japan, a major consumer of CFC solvents for electronics manufacturing, was leery of changes that might raise the cost of doing business. Now Matsushita, NEC and Sony all have programs to eliminate the use of CFCs by 1995, five years in advance of the protocol deadline.

While there has been some backpedaling at the highest levels of the Bush administration, U.S. corporations are taking the initiative in getting rid of their ozone-reducing chemicals. The Hughes Corporation now uses a chemical derived from lemon juice (yes, lemon juice) instead of CFCs in its weapons-manufacturing program. Northern Telecom, a Canadian firm that does most of its business in the United States, has developed soldering processes that

do not need cleaning and has thus become the first major North American company to end reliance on CFCs throughout its operations. "Business is moving faster than the laws require," says Stephen Andersen, an Environmental Protection Agency official who cochairs a Montreal Protocol assessment panel. "They're finding they can save money and improve performance."

*W*hile there has been some backpedaling at the highest levels of the Bush administration, U.S. corporations are taking the initiative in getting rid of their ozone-reducing chemicals.

One uniquely American problem—the 82 million U.S. cars equipped with air conditioners—inspired an enterprising solution. Some automobile mechanics found a patented but uncommercialized machine that enables repair shops to recycle CFC-12 from auto air conditioners rather than vent it into the air. Then they persuaded the Big Three U.S. automakers to require company-owned service centers to install the new device. As a result, 160,000 of these machines had been sold as of January 1992. "The quicker we get out of these CFCs, the better off we're going to be," says Simon Oulouhojian, a service-station owner in Upper Darby, Pennsylvania. "We've got kids, too."

Mexico and Thailand have announced that they would like to phase out CFCs on the same timetable as the developed nations. One factor spurring them on may be the likelihood that exports not meeting strict ozone-friendly standards could soon face international sanctions. But there is also grassroots pressure in some developing countries. In Mexico, for example, consumer complaints persuaded local manufacturers that it was time to begin removing CFCs from aerosol products. The changeover happened so quickly that when one company ran out of labels saying "This is a CFC-free Product," store managers rejected the shipment, knowing that many of their customers would leave unlabeled spray cans on the shelf.

Behind the Ozone Curtain

The countries of Eastern Europe and the former Soviet Union have tougher problems. Faced with a collapsing economy, rising crime and open fighting among its members, the new Commonwealth of Independent States has pushed environmental issues far down on its list of priorities. The Russian people show no special interest in the ozone problem. Whatever aerosol cans and foam products make it to market in Moscow these days are immediately snapped up by buyers who either do not know about CFCs or do not particularly care.

In Czechoslovakia and Poland, most households have CFC-based refrigerators and others badly want them. Neither country has put in place a system for recovering the coolants. Says an official at the Ministry of Environmental Protection in Warsaw: "If we are not able to solve the problem of disposal of used bottles, plastic items and batteries, what can we say about the proper disposal of refrigerators?"

The task is also daunting in the rapidly developing countries of China and India. Together they now contribute 3 percent of the world's burden of ozone-depleting chemicals, but their potential demand for CFC products is so great that without the cooperation of both countries, any plan to heal the ozone hole is destined to fail. China's 800 million consumers, encouraged by more than ten years of economic reform, are ravenous for luxury items such as aerosol cosmetics and air conditioners, and Chinese industry cannot make them fast enough. In the early 1980s China produced 500,000 refrigerators a year; now it churns out some 8 million annually. The Chinese environmental protection agency says it wants the country to switch to non-CFC technologies but does not have the authority to make industry do so.

India, which in the early 1970s invested heavily in the purchase of Western refrigeration technology, today not only manufactures its own refrigerators but exports CFC compressors. Says Ashish Kothari of Kalpavriksh, India's best-known environmental group. "Our development strategies cannot be sacrificed for the destruction of the environment caused by the West." And then there is the cost of changing technologies. "India recognizes the threat to the environment and the necessity for a global burden sharing to control it," says Maneka Gandhi, former Minister of the Environment, who represented India at the Montreal Protocol negotiations. "But is it fair that the industrialized countries who are responsible for the ozone depletion should arm-twist the poorer nations into bearing the cost of their mistakes?"

Both India and China refused to sign the original Montreal Protocol, but they were placated by the creation in 1990 of a special $240 million fund, financed by the developed countries, to help developing nations switch to CFC-free technologies. China signed the revised protocol last year, and India now expects to follow suit. The United States initially balked at the idea of ozone-linked foreign aid but agreed to put up 25 percent of the money after language was added to the agreement stipulating that American willingness to help countries pay for CFC phaseouts would not be taken as a precedent in solving other environmental problems.

Europe, Japan and the United States still need to set up a large, separate fund to help the former Soviet Union and other East European countries wean themselves from CFCs. That will be difficult to do during hard economic times. But what is the alternative? What price is too high to protect the irreplaceable atmosphere shared by East and West, by South and North?

 EARTH CARE ACTION

In Search of a Magic Bullet

Why can't technology rescue the world from the mess that technology created? Isn't there a quick fix? Scientists know there isn't, but that doesn't stop them from musing about fanciful schemes for mechanically or chemically refurbishing the ozone layer in short order. By discussing and critiquing these ideas, researchers hope to educate the public about the dangers of climate engineering as well as learn for themselves the feasibility of various solutions.

"One of the common suggestions is, 'Why don't we just ship L.A.'s ozone up?'" says chemist Sherwood Rowland. "Well, 30 percent of the ozone is in the stratosphere, and it drifts down from there to the lower atmosphere rather than the other way around. The energy that would be needed to move the ozone up is about 2½ times all of our current global power use. If you could take every power plant in the world, every piece of coal and every oil tanker, the energy would be insufficient—and then you'd still have the problem of how to get the ozone up there."

Isn't there a quick fix? Scientists know there isn't, but that doesn't stop them from musing about fanciful schemes.

Considering that there are 350 million tons of ozone in the stratosphere, it would take 350,000 trips by specially outfitted 747 freighters, which can carry 100 tons of cargo, to replace even a tenth of the protective gas. Alternatively, climate engineers could shoot multi-ton bullets made up of frozen ozone into the upper reaches of atmosphere. But the technology for designing and building the tens of thousands of big guns that would be required does not yet exist—not to mention the fact that compressed ozone is dangerously explosive. Furthermore, neither of these solutions attacks the heart of the problem, those long-lived CFCs, which would break down any replacement ozone as well.

As a result, some researchers are focusing their attention on the culprit molecules rather than the victims. Atmospheric scientists Richard Turco of the University of California, Los Angeles, and Ralph Cicerone of the University of California, Irvine, are exploring the idea of injecting into the stratosphere two chemicals—propane and ethane—that would combine with CFCs to produce an extremely weak, and therefore environmentally safe, solution of hydrochloric acid. That strategy would interrupt the CFCs' 100-year destruction cycle, and has the further advantage of requiring only 1,000 jumbo-jet flights over a single, critical 30-day period every year for the next several decades. The products involved are readily available. However, in order for the process to work efficiently, these chemicals must reach from 15 to 25 km (9 to 15 miles) above the earth, and airplanes cannot fly through that entire range. Moreover, the researchers calcu-

> ## WHAT YOU CAN DO
>
> Do not buy foam-construction insulation, cushions, food containers or packaging chips unless you know they are manufactured without ozone-damaging chemicals.

late, there is a chance the plan could backfire and accelerate ozone depletion.

At Princeton University, physicist Thomas Stix has suggested using lasers to blast the CFCs out of the air before they can reach the stratosphere and attack the ozone. His idea is to tune the lasers to a series of wavelengths so that only the offensive molecules would be destroyed. Admittedly, the energy requirement would still be exorbitant, but Stix believes that a 20-fold improvement in the overall efficiency of this approach could make it feasible. Even so, tens of thousands of lasers would have to be designed, tested and built before the first CFC molecule could get zapped. If this is the best idea for reviving the ozone layer, an ounce of prevention is worth more than many tons of cure.

PESTICIDES

Children and lawn chemicals—should they mix? Contrary to what the lawn-care industry has been telling us for years, pesticides are poison and may present a very real hazard to human health. (Sally Shenk Ullman)

GETTING AT THE ROOTS OF A NATIONAL OBSESSION

By Ted Gup

For years, Bonnie Kroll paid a lawn-care company to spray her yard regularly with pest-killing chemicals. Her goal: a carpet of perfect, emerald-green grass. That was until eight years ago, when her black Labrador retriever, Nera, died shortly after one of the treatments. "My parents let the dog out on the wet lawn, the dog rolled in it—end of dog," recalls Kroll. She later had the animal autopsied and spent $4,000 investigating its death. The cause: acute kidney failure, which evidence seemed to show was the result of toxic chemical poisoning.

Since then, the former suburban Philadelphia borough president has been on a crusade to make the lawn-pesticide industry more accountable to the public. "I can't beat the system or the chemical companies," she says. "I can only keep alerting people to the dangers."

Kroll's efforts to get the word out have been relentless. She has appeared on television, placed ads in the newspaper and testified before the state agriculture department. She got the local school board to post warning signs on athletic fields treated with pesticides. And she convinced state lawmakers to introduce bills that would force lawn-care companies to provide more detailed information to consumers about chemical treatments.

"I'm like a reformed smoker," she says. "I used a lawn-spraying service for years. Now I look at it in retrospect and think how sad that I thought it was so important to have this beautiful green lawn."

Toxic Green

Kroll is hardly alone in her disenchantment with lawn pesticides. She's part of a growing nationwide rebellion against what many describe as the indiscriminate use of toxic chemicals in lawn care. Increasingly, Americans are demanding that their communities reassess the risks involved with unnecessary use of certain pest-killing chemicals.

"Poisoning the landscape is always a dangerous idea. Poisoning the landscape to have a pretty lawn is an absurd idea," says National Wildlife Federation toxicologist Jerry Poje. "People who bought into this failed strategy are now recognizing what a terrible idea it is."

The traditional reliance on synthetic lawn pesticide—a $1.5-billion-plus industry—is deeply rooted. Each year, according to a 1990 Congressional report, Americans shake, dab or spray some 67 million pounds of chemicals on their lawns to eradicate fungus, insects and weeds. Activists struggling to break the nation of its lawn-chemical dependency have taken their fight into the courtroom, city hall and Congress. Evidently, their message is getting through. Around the country, numerous laws imposing new restrictions on lawn-care companies have either been passed in recent years or are under consideration.

---

*P*oisoning the landscape is always a dangerous idea. Poisoning the landscape to have a pretty lawn is an absurd idea.

Some states, including Maine, Maryland, New Jersey and Rhode Island, now require warning signs on treated property. Others demand that companies alert consumers to the name of the pesticide, the amount used and the target pest.

The lawn-care industry insists its products are safe when used as directed and that they undergo rigorous scrutiny by the Environmental Protection Agency (EPA). Last year [1990], Warren Stickle, president of the Chemical Producers and Distributors Association, told a Congressional subcom-

mittee investigating pesticides that "most of these products have a long record of common usage, with a safety record of no adverse health effects."

Nevertheless, public concern is on the rise. What many people find disturbing about the nation's love of lawn pesticides is the widespread tendency among Americans to apply the chemicals in huge amounts—and often when they're not necessary at all. "You rarely need synthetic pesticides, particularly in regard to maintaining lawn," says Sheila Daar, director of the Bio-Integral Resource Center, a biological-control resource laboratory in Berkeley, California. "There is always another way that's environmentally sound."

Human Costs

The rise of the lawn-care industry in this country two decades ago owes its success to several factors: the explosion in the suburban housing market, the advent of the dual-income family, with the premium it placed on homeowners' time, and the profusion of chemical pesticides that emerged after World War II.

Like the power lawn mower, chemicals promised to take the toil out of yard maintenance. From the mid-1960s well into the 1980s, the lawn-care industry grew at a rate as high as 25 percent per year. Now it is facing its stiffest challenge. The rebellion against lawn pesticides is especially evident in New York State. In Albany, thousands of citizens calling themselves the New York Coalition for Alternatives to Pesticides have taken their campaign to neighborhoods, schools, corporations, cities and state agencies—to anyone who will listen to their warnings about pesticides and prescriptions for natural alternatives. The coalition recently helped convince municipal authorities in Buffalo to form a commission to find the least toxic means to deal with lawn pests.

Much of the campaign against chemical pesticides, however, is being waged by individuals. Take Sue Evans, a systems analyst with General Electric's Research and Development Center in Schenectady. For years she petitioned GE officials to reduce the amount of chemicals they used to maintain the center's 90 acres of lawns. Last year, the company responded by converting nearly a third of the property into a pesticide-free wildlife habitat, complete with nature trails and nesting sites for bluebirds.

At Cornell University in Ithaca, sophomore Yona Brown became concerned when fellow students complained of itchy eyes and headaches following a pesticide application on campus. So last fall, Brown helped organize a petition drive to halt the application of pesticides used for purely cosmetic purposes. Hal Craft, Cornell's vice president for facilities, says the school is now seeking ways to "drastically reduce" pesticide use.

One of the most widely used and most controversial of lawn chemicals is known simply as 2,4-D. About four million pounds of the herbicide are applied in the country each year. A 1986 study by the National Cancer Institute found that Kansas wheat farmers who frequently applied 2,4-D experienced a marked increase in the incidence of non-Hodgkin's lymphoma.

Such findings persuaded the nation's largest lawn-care company, ChemLawn, to suspend use of 2,4-D after relying on the product for many years, says a company official. However, a consortium of 2,4-D producers continues to challenge the connection between the pesticide and cancer, citing studies in New Zealand and elsewhere that failed to find such a link.

Animal Costs

Another common pesticide that has gained unwanted attention is diazinon, widely used on a variety of insect pests in gardens and on agricultural and ornamental nursery sites. In Hempstead, New York, 700 Atlantic brant geese died of acute diazinon poisoning after three golf course fairways were treated with the insecticide in 1984.

It was not an isolated incident. Around the same time, the EPA received 60 reports of bird kills from 18 states, affecting 23 species. Of those incidents, 20 occurred on golf courses. In 1986, the EPA banned the use of diazinon on golf courses and sod farms. (Because the bird fatalities were mostly associated with large, grassy expanses, the EPA still allows diazinon's use on residential lawns, says an agency spokesman.)

Ciba-Geigy, the primary producer of diazinon, has challenged the ban. Everett Cowett, the company's director of technical services, acknowledges diazinon can be hazardous to birds. But, he says, it remains one of the most effective insecticides available, and that "the alternatives are equally dangerous" to birds. Avian casualties must be put into perspective, he says: "Automobiles run over more birds than diazinon kills."

Despite industry assurances, critics question both the safety and efficacy of chemical pesticides. "These poison purveyors **never give** a full accounting of the risks," says the National Wildlife Federation's Jerry Poje. "People and the environment have been contaminated far too often for us to accept industry's word on safety." Poje and other scientists charge that EPA's reviews of the products are woefully inadequate. The agency lacks critical details about long-term safety and alternative approaches to pest management, they say, arguing it relies on data provided by the very industry that produces the chemicals.

The Public Trust

The EPA says it cannot afford to conduct independent tests—which can cost as much as $30 million per product—but that the agency monitors industry results. That, however, does little to quiet some opponents. "The EPA has failed the public's confidence," charges Jay Feldman, president of the National Coalition against the Misuse of Pesticide. "Does a green lawn justify exposure to carcinogens, neurotoxins and chemicals that can cause birth defects, genetic damage and toxic sensitization? We do not think so."

One witness at last year's Congressional hearing, Thomas Prior, told how his 30-year-old brother, Navy Lieutenant George Prior, died in 1982 after playing golf on a course that had recently been treated with a fungicide: "He became grotesquely swollen. Enormous blisters appeared on his body. One by one his organs failed, his skin sloughed off and he became blind." Fourteen days later he died. According to the EPA, a private lawn-care company reached an out-of-court settlement in the case, though the agency lacks sufficient evidence to conclude pesticides killed Lieutenant Prior.

Reports of fatalities from acute toxicity are very rare. More common are poisoned pets and hypersensitive reactions in humans. Eleven-year-old Kevin Ryan of Arlington Heights, Illinois, told the subcommittee of his symptoms when exposed to pesticides: numbing in his arms and hands, achiness in his joints, pressure on his chest, difficulty breathing, nausea, depression. "I can't even play in my own backyard because my neighbors spray their lawns and trees," said Ryan.

*D*oes a green lawn justify exposure to carcinogens, neurotoxins and chemicals that can cause birth defects, genetic damage and toxic sensitization?

Many others share his dilemma. Two years ago [1989], for instance, a doctor determined that Marian Arminger's eight-year-old son, Jared, was hypersensitive to chemicals, particularly lawn-care products. Every time the boy comes in contact with them he succumbs to a variety of symptoms, ranging from diarrhea and excessive salivation to swollen glands and a racing heartbeat. Today the Baltimore, Maryland, family is listed on a state registry of persons certified by a physician to be acutely sensitive to pesticides. By law, lawn-care companies treating adjacent properties must warn the family in advance.

Heightened concern about pesticides is redefining the lawn-care market. "The picture-perfect, green, weed-free lawn is a myth, a marketing lie—it doesn't exist," says Phil Catron, once regional technical manager for ChemLawn. Today he has his own company, NaturaLawn, based in Bethesda, Maryland.

Catron maintains that the routine use of synthetics by lawn-care companies can destroy a lawn. "Over a period of two or three seasons," he says, "the [chemically treated] lawns would generally have a weakened root system and die from stress."

Alternatives Arise

Insecticides kill not only unwanted insects, says Catron, but also beneficial ones vital for aerating the soil. Instead of treating specific problem areas, many people douse the entire lawn, whether it needs treatment or not. "If I cut my arm," he points out, "I put a bandage on my cut. I don't wrap my entire body." This kind of overkill can backfire. As Sheila Daar explains, a heavy dose of chemical pesticides often wipes out a pest species' natural enemies. "And when you kill the 'beneficials,' " she points out, "you increase the number of pests that you have to control."

When Phil Catron declares war on whit grubs (U-shaped worms that devour the roots of grass), he turns to naturally occurring microorganisms called milky spore bacteria. Infected with milky spore, the grubs turn milky white and die, releasing more bacteria into the soil.

Milky spore is the key ingredient in Grub Attack, a product of the Ringer Corporation based in Minneapolis. The 30-year-old competitor, Safer Inc., also markets a natural insecticide derived from chrysanthemums as well as pesticides based on plant and animal fats—the sort used in household soaps.

The appeal of less toxic approaches is rapidly gaining favor even among land managers who once relied heavily on chemicals. "We have made a very conscious effort to reduce the amount of pesticides we use," says Sylvester March, chief horticulturist for the 450-acre U.S. National Arboretum in Washington, D.C. Pro golfer Arnold Palmer, who owns and manages some 20 golf courses, has put out the word that he wants fewer chemicals used on his fairways and greens.

Within the past few years, the National Park Service has sharply cut back on its use of chemical pesticides. In 1983, for example, the service used 32,000 pounds of chemicals to treat 15,000 acres of federal park land. In 1986 (the most recent year for which figures are available), the agency used only 25,000 pounds to treat 51,000 acres. The service is also turning away from regularly scheduled applications of lawn pesticides. It now relies more on nonchemical solutions and monitoring of pest population levels.

The lawn-care industry itself has begun promoting a more "enlightened" attitude—even as it defends its standard line of products. Some companies show restraint in their use of chemicals, focusing more on spot applications than blanket treatments.

However, many in the industry express skepticism about nonchemical products. Thomas J. Delaney, of the Professional Lawn Care Association of America, told the subcommittee that "effectiveness of currently available biological control methods remains disappointing and unacceptable to our customers."

Indeed, natural pesticides generally require more work and do not produce results as fast as synthetic chemicals. But proponents say the alternatives in many cases kill only target pests, break down harmlessly in the soil and are less likely to create resistance in insects.

ChemLies

Not long ago, broad safety claims by lawn-care giant ChemLawn, which boasts more than 1.5 million residential customers, landed that company in hot water with the state of New York. In 1988, the Ohio-based company was sued by the New York attorney general for allegedly making false and misleading statements. One such claim: " A child would have to swallow the amount of pesticide found in almost ten cups of treated lawn clippings to equal the toxicity of one baby aspirin."

Last summer [1990], the company agreed not to make a number of product-safety statements that the state found unsupported. Says assistant attorney general Ann Goldweber, "There aren't enough data available to assure us that these pesticides are safe. ChemLawn uses the lack of information in an affirmative way to say, 'Because there are no data, these products are safe.' "

In its annual report for 1989—a "very tough year"—ChemLawn disclosed that an increased cancellation rate reduced the number of its residential customers by 6 percent. The company closed some branches and laid off more than 500 employees. ChemLawn officials blame the sagging profits on changes in New York's laws, among other things, but acknowledge that heightened sensitivity to the use of chemicals also played a part. Says Roger Yeary, "I think, in hindsight, we'd probably change the company's name." (See Editor's Note below.)

ChemLawn uses the lack of information in an affirmative way to say, "Because there are no data, these products are safe."

It will take more than a name change to address the concerns of people like Bonnie Kroll and Marian Arminger. "If there were a place to move to get away from pesticides, many people would move," says Arminger. "But there's no place to go."

(Editor's Note: While ChemLawn has not changed its name, that name now appears in only small print on many of the company's vehicles. The firm's pesticide-carrying trucks now have the word "Eco-Scape" emblazoned in bold letters on their sides.)

From *National Wildlife,* June/July 1991. Reprinted by permission.

How to Maintain a Chemical-Free Lawn

By Ted Gup

Before declaring chemical warfare on your lawn, consider whether you have taken all possible steps to make your property inhospitable to pests. A great deal can be done to deter pests, much of it going back to the elemental roots of gardening. Following are a few of the basic considerations:

Provide healthy soil. A nutrient-rich layer of topsoil, five to six inches deep, is a critical first step. If your yard has less than this, consider adding composted manure (free of weed seeds) across an existing lawn, but not so much as to smother it. In addition to organic compost, the Environmental Protection Agency (EPA) suggests grass clippings, bone meal, cottonseed meal or dried blood, which release vital nutrients into the soil without destroying natural microorganisms and earthworms vital to its well-being. Warns the EPA: "Quick-release chemical fertilizers can brown the grass, induce pest infestations, increase thatch buildup and promote leaf growth at the expense of healthy root growth, leaving the grass susceptible to summer heat, drought, disease and compaction."

Test the soil for any imbalance in alkalinity or acidity—the so-called pH balance. If it is excessively acidic, add lime. If too alkaline, add sulphur and perhaps gypsum, depending upon your region. The tendency is toward acidity in the East and alkalinity in the West. Consult your county's cooperative extension office for information about your area.

Use care in selecting grasses. Plant grass varieties that are right for your particular climate and condition. According to the National Coalition against the Misuse of Pesticides (NCAMP), a mix of grasses offers better protection against pests and disease than a pure stand. Because of profound regional variations, ask your county extension office for the optimum mix. In some instances, planting additional seeds among established lawns can reduce weed infestation and repair damage.

Maintain the lawn properly. Always make sure that your mower's blade is sharp. A dull blade leaves an uneven cut that can retard grass recovery and expose grass to loss of moisture. Never give your lawn a butch cut. Grass that is cut too short becomes more vulnerable to stress from heat and drought. Cutting the grass too short also allows too much light to get to the soil, enabling weeds—especially crabgrass—to prosper. NCAMP recommends setting the mower blades as high as possible, though of course the proper height depends on the variety of grass. To avoid unnecessary stress to the lawn, do not mow in the heat of the day. Prevent excessive buildup of thatch, the grass stems and roots that accumulate just below the surface and can promote pest infestation.

Depending upon the region and character of the lawn, the EPA suggests applying organic matter or gypsum, then watering it to help release the gypsum. This helps reduce alkalinity and make a better growing medium for turf. Avoid overwatering the lawn, as excessive moisture can lead to shallow root systems and disease-prone lawns. Where soils are compacted, aeration (the removal of cores of soil) may be helpful. Aeration allows air, water and nutrients to seep in, promoting healthy root growth. The best natural aerators are ants and earthworms—often the first victims of pesticides.

Monitor for pests. The most important step in effective lawn maintenance is watching regularly for trouble to keep small problems from becoming big ones.

Check the lawn at least once a week, keeping an eye out for signs of insect, fungus or weed infestation. Weeds can be removed by hand and controlled to some extent by mowing. If an herbicide is called for, select one that is least toxic, such as a fatty-acid soap. If the problem is insects, first identify the species. For the white grub or the Japanese beetle, NCAMP recommends milky spore bacteria, a naturally occurring insect disease. For chinch bugs, try using a wash of insecticidal soap over the thatch.

Sheila Daar, director of the Bio-Integral Resource Center in Berkeley, California, offers some additional tips on how to recognize and deal with yard pests. In many parts of the country, she explains, insects don't become ac-

tive until the soil temperature rises above 50°F. She suggests that homeowners buy a soil thermometer and stick it into the ground next to the sidewalk for easy access. When the soil temperature hits 50°, "start looking for suspicious browning of the lawn or other signs of trouble."

If you see a suspicious spot, mix a couple of tablespoons of detergent in a gallon of water and pour the concoction over the spot. "This will force the pests up to the surface where you can count them," says Daar. If the problem is white grubs, cut a flap in the lawn and count the larvae. "It's a numbers game," she says. "The lawn can tolerate a certain number of pests. It's all a question of how healthy the lawn is."

Homeowners struggling to keep up with the Joneses might keep in mind that, in general, "the most heavily managed lawns—golf courses, for example—are usually the ones with the biggest pest problems," says Daar. Applications of chemical pesticides kill off susceptible organisms, often starting with the beneficial ones, she says. Pest species, however, often develop resistance and continue to reproduce, requiring more frequent applications, higher doses and stronger chemicals—a cycle called the pesticide treadmill.

The Bio-Integral Resource Center publishes brochures with tips on lawn care and pest control. For a copy of the "Least Toxic Pest Management Publications Catalog," send $1 to P.O. Box 7414, Berkeley, CA 94707.

From *National Wildlife*, June/July 1991. Reprinted by permission.

Sermon on the Farm

By Susan Katz Miller

Tramping through papaya fields, praising the virtues of pig manure, the man with the baby monkey on his arm rushes ahead of his visitors. Unfazed by the heat and humidity of West Africa's rainy season, he gestures proudly toward the flourishing vegetable gardens, the populous fish ponds, the pens of freckled guinea fowl. Yet Nzamujo Ugwubelam is more than an enthusiastic farmer. Dominican priest and computer engineer, he is also an apostle for an environmentally sound agriculture tuned to the needs of African farmers.

Ten years ago, the Nigerian-born Father Nzamujo abandoned a comfortable life teaching in southern California and moved to the West African country of Benin. There he transformed 25 acres of bush into the Songhai Project: a model farm, research station and training ground for young farmers. "I wanted to know if we could raise the standard of living of Africans by harnessing the resources we have on this continent," he recalls. "So I set out to design an ecologically balanced, indigenous, self-sustaining agricultural system."

Early signs indicate that this charismatic man is succeeding in helping African farmers develop alternatives to Western-style agri-bigness. As such, Ugwubelam and Songhai are in the vanguard of a worldwide movement to make agriculture "sustainable," a code word for environmentally benign farming that saves the land by using fewer costly inputs such as store-bought fertilizers, pesticides and complex machinery.

Also called "low-input" farming, the approach is one answer to a web of troubles afflicting Africans and other Third World citizens—problems such as deforestation, loss of plant and animal species, food shortages and massive debt. "We're developing new models appropriate to areas without infrastructure, areas with low soil and water resources," sums up Richard Harwood, a sustainable agriculture expert at Michigan State University.

Songhai's founder puts it more colorfully. "The present system is bankrupt ecologically," says the priest. "We must treat the soil as a living woman who must be fed, caressed and given the best you have. Instead, we have become like pimps, treating the land like a prostitute; and then of course it refuses to grow."

A New Approach

With a job teaching at the University of California at Irvine in the early 1980s, Father Nzamujo (who had earlier studied for the priesthood in Nigeria) was well on his way to becoming a successful First World professional. But he could not forget summers spent on his grandmother's farm in Nigeria. "Even when I lived in downtown Los Angeles, I was gardening and raising quails," he says. "People

To Nzamujo Ugwubelam, pictured here with an animal companion, use of costly farm chemicals is "a mortal sin." (Susan Katz Miller)

thought I was crazy." Memory mingled with sorrow after he began seeing TV images of drought and famine in the Sahel, the semi-arid zone of Africa south of the Sahara.

"People told me to just stay quiet, to stay in my nice air-conditioned California office." Instead, Father Nzamujo traveled home to Africa and was shocked by the poverty he saw after more than a quarter-century of independence. "Development experts were trying to graft the African economies onto the Western economy," he says. "I became convinced that they were using the wrong economic equa-

tion." Determined to help, he moved in 1984 to Benin, a narrow wedge of a nation along West Africa's old Slave Coast.

Benin's coast is lined with palm trees planted to produce tropical oils for export— part of the same cash-cropping syndrome that in other West African former colonies led to extensive cultivation of coffee, cocoa, tobacco or peanuts. Unfortunately for Benin's economy, the value of palm oils has plummeted in recent years, as consumers abroad have come to shun oils high in saturated fats.

Cash-crop farming has proved burdensome in other ways. Planting the same crop year after year wears out thin tropical soils. To compensate, agriculture advisers have long urged Africans to spread chemical fertilizers and pesticides. Such thinking was at the heart of the "Green Revolution" of the 1960s and 1970s, when modern agricultural methods were seen as the key to increasing Third World food harvests. But these imported agricultural "inputs" were often beyond the means of African farmers. The price tag included environmental problems such as soil erosion, stream pollution from chemical run-off and poisoning of village wells.

"The Green Revolution was designed in developed countries, and then they tried to transfer it to developing countries," says Stephen Gliessman, director of agroecology at the University of California at Santa Cruz. "They did raise yields, but neither the environmental nor the social impact was assessed beforehand."

Per capita food production has actually declined across sub-Saharan Africa in the last decade. And while the largest aid donors still emphasize big agriculture, there is a growing realization at the grassroots level—down on the African farm—that a new approach is necessary.

Doing without Pesticides

Enter Father Nzamujo and Songhai. Named after a precolonial West African empire, the project aims to provide farmers with alternative techniques that help them achieve "long-term stability and sustainability," says Father Nzamujo, as well as a better "quality of life for our children and grandchildren." Songhai's methods also let farmers use resources of the remaining African forest, rather than clearing it to plant cash crops.

Clearly, something is working at Songhai, a sprawling complex of fields, pens, hutches, ponds and bungalows on the outskirts of Benin's capital, Porto Novo. Daily at dawn, Father Nzamujo says Mass in Songhai's simple chapel. He then makes rounds, greeting the students who are busy tackling other farm chores.

Songhai's system—in which everything is used, nothing discarded—is based on a relationship among fish farming, plant crops and livestock. "The only way you can avoid using chemical fertilizers is to use everything else," says Father Nzamujo, tossing a bucket of pig manure into a pond to feed the *Tilapia nilotica*, a fast-growing fish sold to local villagers. Run-off from fish ponds, rich in nutrients, irrigates gardens of cabbages, tomatoes and onions. Songhai students—those attending workshops or striving toward a two-year diploma—cull and dry small fish from the ponds to make chicken feed.

"The use of chemical pesticides or fertilizers is a mortal sin here—we don't use even a drop," says the priest-turned-farmer as he checks on the progress of three students tending a field of huge lettuces. Doing without pesticides makes Songhai's techniques affordable for African farmers and preserves beneficial bugs and bacteria living in the soil.

The use of chemical pesticides or fertilizers is a mortal sin here—we don't use even a drop.

A key component of any organic farm is planting of complementary species. Leeks at Songhai grow beside eggplants, keeping bugs away with their pungent odor. Tall cassava plants provide shade for low-growing cocoyam.

In Songhai's kitchen, more than food comes from the farm. Stoves run on "biogas," or methane, given off by decomposition of—yet again—pig manure and other organic detritus. "Look how pure and blue that flame is," exclaims Father Nzamujo, striking a match at a tank of captured methane. He hopes that one day widespread use of biogas will reduce the number of trees West African villagers cut for fuel wood, a cause of deforestation.

Big Dreams

One of the priest's biggest concerns is the loss of West African plant and animal species with traditional uses. "Modern agriculture has done a lot of harm by neglecting and suppressing the knowledge of a broad range of native plants and concentrating on corn and soybeans," he says. "I see Songhai as a refuge, a sort of living laboratory of species." He stops to breathe the scent of a basil variety that was becoming rare in Benin. Now Songhai is giving seeds to neighboring farmers.

Two green monkeys are allowed the run of the farm, part of a strategy to identify medicinal plants. By observing which plants the monkeys eat when ill, Songhai staff have identified a deworming medicine that has proved useful for livestock.

The Songhai menagerie keeps growing, with such farmyard pets as a giant land tortoise, a four-foot crocodile and an orphan monkey named "Australopithecus" for her resemblance to man's early ancestors. The little monkey clings to Father Nzamujo's arm as he escorts the steady stream of visitors (they range from European tourists to Nigerian bishops) who have heard of the gospel of alternative agriculture practiced in the Benin bush.

Currently, 60 students are in residence, their tuition and living expenses paid by donations to Songhai and profits from the farm. Last year, 250 Beninese applied for only 23 spaces in the program. Students flock to Songhai because Africa's economic crisis deprives them of other options; even those with university degrees return to the land for want of other work.

Standing under the shade of a mango tree at day's end, Father Nzamujo admits he has big dreams for Songhai and hopes to see his farming philosophy take root throughout West Africa. That will take time, money and a continuing willingness to cross-fertilize African

knowledge with useful technology from the West. "My dream is for Africans to be able to cooperate with Westerners without losing their identity," says the priest. "To survive in the world today, you have to be inoculated with all of the cultures."

And if you can tweak the West while bringing home its best, so much the better. More than a year ago, on the eve of the Persian Gulf War, Father Nzamujo was traveling from California to Benin. In his briefcase were a dozen quail eggs destined for the farm. At a Paris stopover, a nervous airport security official demanded that the priest break open an egg.

"Look, you see, this was a female!" cried Father Nzamujo, after cracking the egg. "Think of all the quails that could have been raised from this one female to feed the Africans." While the embarrassed security man offered to pay for the egg, the priest—with a hint of mischief—says he assured him the egg and its broken promise would stick forever on his conscience. And with that Father Nzamujo took 11 eggs to the farmers of Africa.

From *International Wildlife*, March/April 1992. Reprinted by permission.

Taking Aim
at a Deadly Chemical

By Diana West

The old-timer plunged into the piney thicket of the Slash, a densely wooded tract on Curles Neck Farm in Henrico County, Virginia. Great armies once crisscrossed this quiet land: Union troops sweeping through en route to Richmond; Confederates driving the foe to the banks of the nearby James River. But on a fine April day in 1985, the blood and powder of pitched battle are unimaginable. Only a metal detector, like the one wielded by the old man, can uncover the telltale Minié balls and brass buttons lodged in the forest floor.

As the relic hunter—whom locals knew only as "Mr. Phelps"—trekked deeper into the woods, he glimpsed the white head and dusky wings of a bald eagle, one of the endangered birds that nest in this remote stretch of Tidewater Virginia.

But this eagle was on its back, wings extended, legs outstretched. Dead. Staring down at the carcass, the old man opened his penknife, cut off the raptor's leg and pocketed the bird's aluminum identification band. While not a Civil War souvenir, the metal band might be worth keeping.

Trudging out of the woods that day in 1985, old Mr. Phelps had no idea that his chance discovery would ignite a new war in Virginia, one that would pit agricultural interests against protectors of wildlife, big business

against state government—all over the use of granular carbofuran, a potent insecticidal poison developed in 1970 by the FMC Corporation of Philadelphia.

Last year [1991], Virginia environmentalists declared victory in that war when state regulators, swayed by support from Governor L. Douglas Wilder, voted to ban the sale and use of granular carbofuran in the Old Dominion. For the first time anywhere, a chemical had been banned solely because it imperiled wildlife. Even DDT, notorious for ravaging the nation's eagle, falcon and pelican populations and outlawed in this country in 1972, remained on the market until evidence mounted that it threatened humans as well as animals.

> *The process for canceling pesticides is weighted heavily in favor of the manufacturers—which is backward. It should be weighted in favor of the environment.*

Had it not been for a handful of Virginia conservationists mobilized by the death of the Curles Neck eagle, the carbofuran story might have ended in political impasse. For while Virginians took aim at the poison, the feds daw-dled. By 1989, the Environmental Protection Agency (EPA) had the facts: Carbofuran was killing two million birds every year in this country. Since 1972, carbofuran-poisoned birds have been found in 23 states. Supported by the U.S. Fish and Wildlife Service and groups such as the National Wildlife Federation, the EPA recommended canceling the pesticide's regis-tration—that is, banning it. But nothing hap-pened.

There was no action, says Robert Irvin, a National Wildlife Federation attorney, because "The process for canceling pesticides is weighted heavily in favor of the manufactur-ers—which is backward. It should be weighted in favor of the environment."

"If a chemical is not going to cause a problem for humans," adds John Bascietto, a former EPA biologist, "it generally won't be regulated for wildlife concerns."

A Pinky-Purple Insecticide

Ultimately, Virginia would get its ban and the federal government would be stirred into action, with EPA crafting an agreement with FMC designed to eliminate granular car-bofuran's use throughout the United States. But in 1985, when the dead eagle turned up at Curles Neck, victory was a long way off.

"Our position on this thing right from the start was: Forget EPA. They're sitting on their hands," says Robert Duncan, chief of the wild-life division at the Virginia Department of Game and Inland Fisheries. One of the band that pushed to cancel carbofuran, Duncan says, "This was a Virginia problem, one that was documented in Virginia, and Virginia was

going to take care of it—i.e., get rid of it. But that was more easily said than done."

Carbofuran is the name of the poison in Furadan 15G, a pinky-purple insecticide that farmers throughout the United States sow by the sackful to protect corn, rice and other crops from root-eating nematodes and insects. It is deadly stuff. One of carbofuran's ingredients is methyl isocyanate, the compound that killed some 2,000 people in 1984 when a gas cloud escaped from a Union Carbide plant in Bhopal, India. Although agricultural carbofuran poses no threat to people, a single granule can kill a songbird.

U.S. Fish and Wildlife Service special agent Don Patterson had scarcely heard of the stuff when he was called in to investigate the Curles Neck eagle's death. As he followed old Mr. Phelps (who had told authorities of his find), he could smell the eagle before he saw it, rapidly decaying in the springtime warmth near the base of its nesting tree. Inside the nest, 60 feet above the ground, he found a dead eaglet. He also found the remains of pigeons and blackbirds, evidence the eagles had proba-bly died from eating poisoned carrion. Patter-son shipped the eagle to a pathology lab for a necropsy.

The verdict, "death by carbofuran," rang a bell. Patterson remembered that Virginia wildlife officials recently had picked up a sick female eagle on the south side of the James River. The sick raptor could barely stand up, let alone fly.

Officials ferried the eagle in a large dog carrier to the Wildlife Center of Virginia, a private veterinary hospital and research center in the Shenandoah Valley. Founded by conser-vationist Ed Clark and veterinarian Stuart Por-ter, the Wildlife Center treats nearly 2,000 wild animals a year, providing researchers an opportunity to monitor the region's environmental problems at close range. Treated and released, the eagle had given Virginia its first warning of carbofuran poisoning.

After reading the necropsy report on the state's second known carbofuran victim—the Curles Neck eagle—Don Patterson alerted farm owner Richard Watkins. "We stopped using Furadan right away," says Watkins, who hasn't used insecticides on his 400 acres of corn at Curles Neck since hearing the news. "We haven't noticed any change in our yield," he adds.

Stopping a Killer

But there were hundreds of thousands of acres of Virginia farmland where Furadan remained in use. Patterson sent copies of the necropsy report to Robert Duncan and Ed Clark, two men he thought could help. "In our wonderful naivete," says Clark, a career conservationist soon to be appointed to the Virginia Council on the Environment, an advisory board, "we thought now that we knew something was poisoning eagles, somebody will do something about it." But stopping a killer chemical would prove much harder than expected.

Spring planting season arrived, and Patterson picked up four more poisoned eagles. By now, Furadan's maker acknowledged a problem with wildlife. But FMC faulted farmers, accusing them of misusing the poison to bait predators, or of mishandling it by spilling it. "We [sponsored] an extensive stewardship program to make sure applicators and growers were sensitive to the risks," FMC spokesman Jeffrey Jacoby says of company efforts in Virginia. "We all certainly didn't want to kill any bald eagles."

Patterson once guided some FMC repre-

sentatives through Virginia eagle country. "While there, we saw 12 [living] eagles," he says. "And one [FMC representative's] comment that came to me, was, 'Hell, we're always going to have a problem here; there's too many eagles!' " Patterson bursts into a wheezy guffaw at the memory.

He also tells of a Virginia farm family that awoke one morning several years ago to find their yard sprinkled with the vivid yellow carcasses of between 50 and 70 goldfinches. "Cats and dogs were already carrying them off," says Patterson, who arrived after carcasses had been collected. Surrounding corn fields, it turned out, had been laced with Furadan.

But it was not until 1990 that officials learned of the event that, in Virginia, would mark the beginning of the end for Furadan. That year, a massive bird kill occurred in eastern Virginia near the Rappahannock River. Agents combing corn fields picked up more than 200 dead sparrows, eastern bluebirds, red-winged blackbirds, blue jays, goldfinches, starlings and other species. "It was right quiet," recalls Bob Duncan, describing the macabre afternoon he visited the site.

The aftermath was anything but quiet. The bird die-off thrust the carbofuran issue before a new state Pesticide Control Board, a ten-member regulatory panel that had come into being a year earlier. The board scheduled public hearings throughout the summer and fall of 1990. Both sides of the pesticide war clashed, noisy and clear. When the last verbal shots had been fired, the board decided to give the manufacturer one more chance.

FMC insisted new safety measures could reduce the risk to birds. The pesticide board told the company to put those measures to work during the 1991 planting season in a 32-county region east of Interstate 95—the heart of Virginia eagle country. The board also asked Duncan's state wildlife division to monitor the measure's effectiveness.

Come spring, FMC launched a publicity blitz. A soaring eagle glided across billboards and baseball caps that FMC distributed across Virginia, tools in a campaign urging customers to "Protect as you plant." The company drew up new instructions for Furadan's use, prescribing smaller doses than previously suggested and improving application techniques.

For all of FMC's noisy drills and maneuvers, the real battle was a quiet one, bird against poison, waged in the freshly planted fields of Virginia. State government monitors would determine the outcome by counting casualties.

"We were a little nervous going into it," says Elizabeth R. Stinson, the state game department biologist who devised the monitoring program. Finding a bird as big as an eagle is hard enough. Spotting a small Savannah sparrow in a vast farm field can be next to impossible. Stinson helped train a team of ten monitors to search corn fields on 11 farms. Through tests, she established that monitors could find from 30 to 50 percent of all dead birds.

Monitors combed fields before and after Furadan's application. As the toll mounted, carcasses of kestrels, robins, water pipits, cowbirds, meadowlarks, cardinals and other birds were sent to labs to confirm the cause of death.

On 900 acres, Stinson's team gathered 62 bird carcasses, 10 ailing birds and 47 "feather spots"—piles of feathers that indicated a bird had been scavenged. Assuming that monitors were, at best, picking up half the carcasses to be found, and considering the hundreds of thousands of acres of Furadan-treated farmland across the state, the monitoring program indicated tens of thousands of birds were dying of pesticide poisoning every spring.

Dirty Tricks

In May 1991, the Pesticide Control Board deliberated on the results of the monitoring program. Virginia Governor Wilder threw his support behind a Furadan ban. "We thought we had it," says Ed Clark, who went home feeling relieved. News headlines one day before the board would vote changed his feelings.

In what Clark calls a "public relations masterpiece," FMC officials announced their own 11th-hour decision to stop selling Furadan in the state of Virginia. Pesticide Control Board chairman George Gilliam accepted the offer, saying a voluntary withdrawal was as good as an official ban.

But what FMC could withdraw, FMC could reintroduce. Besides, even if Furadan 15G stayed off the shelf in Virginia, what was to prevent a farmer from hopping the border to buy pesticides?

"Those of us who knew what was going on behind the scenes went ballistic," says Clark. Impassioned calls went to the governor's office, the state agriculture department, the media. "My phone was melted by the end of the day," Clark says of the blitz.

Ultimately, the pesticide board accepted FMC's offer—but also imposed its own emergency ban on Furadan 15G, beginning June 1, 1991. "Sort of a belts-and-suspenders approach," quips chairman Gilliam, who ended up favoring a ban. Gilliam notes that the state pesticide board banned Furadan after focusing on the issue only one year, while "the EPA has been on this issue since 1985 . . . . I feel if we hadn't had the backbone to act, the EPA may very well have taken another several years."

Instead, EPA took just days to announce

it had reached a multiphase agreement with FMC. ("It took Virginia a year to do something and they only had to worry about one state," an EPA official said of the years of agency deliberation that preceded the agreement. "We had to think about the whole nation.")

The EPA agreement stipulates that FMC will take Furadan 15G off the U.S. market in crop-by-crop stages, beginning with corn. By September 1994, only 2,500 pounds may be sold annually, a thimbleful compared to an estimated ten million pounds that crossed the counter each year.

As some people see it, the long-awaited agreement contains loopholes. For one thing, FMC retains the right to appeal to the EPA before each stage of the phaseout kicks in. Also, a U.S. phaseout by itself provides incomplete protection.

"Many of the same birds we were seeking to protect here are going to be exposed to this chemical in other regions as they migrate," Ed Clark says. In fact, Tarry Lalonde, manager of a West Virginia plant that makes Furadan 15G, told the *Charleston Sunday Gazette-Mail* that exports of the insecticide will actually increase to balance shrinking domestic sales.

And those exports will not be tainted by the stigma of an EPA ban. "If something is banned in the United States," explains former EPA biologist John Bascietto, "it becomes a

---

WHAT YOU CAN DO

If you're tired of mowing, watering and fertilizing lawns—and you don't want to spray for pests—there is an alternative. It's called xeriscaping. The xeriscape alternative is to use paving stones, brick and mulched beds in combination with ground covers, shrubs, plants and trees that thrive naturally in your region.

---

serious problem for overseas regulators not to do anything about it. If it's not completely banned, then foreign regulators are under less pressure to look at it more closely."

But thanks to Virginia conservationists, FMC cannot tell customers that Furadan 15G isn't banned *anywhere.* "Virginians were well served," Robert Duncan says of the battle to banish the bird-killing chemical from the Old Dominion. "And the Pesticide Control Board members deserve credit for that," he drawls, "even if they had to be persuaded to bring it about."

From *National Wildlife,* June/July 1992. Reprinted by permission.

TROPICAL FORESTS

White-handed gibbons sit high atop a Malaysian rain forest. Do they realize that their home is endangered? (Rainforest Action Network)

A MAN WHO WOULD SAVE THE WORLD

By Hank Whittemore

From the city of Belem at the mouth of the Amazon River in Brazil, the flight inland proceeds over vast devastation caused by cattle ranchers, gold miners and loggers. It's three hours to Rendencao, the farthest outpost on the tropical frontier. Then a tiny plane continues until, mercifully, the scene below is transformed into a canopy of lush green treetops shielding perhaps half the plant and animal species on earth.

Later the pilot dives and banks over a clearing of red dirt bordered by small thatch homes—a signal to the Kayapo people of Aukre. When the plane drops amid tall trees to find a thin landing strip and rolls to a stop, scores of villagers emerge staring in silence. Their bodies and faces are painted with intricate designs; they wear colorful bracelets and necklaces of beads. Some of the men carry guns or knives or bows—they are warriors, with a heritage of fierce pride that is centuries old.

The visitor is led into the main yard of the village, where Chief Paiakan stands near the Men's Hut at the center. He is about 37, but the Kayapo do not measure time that way, so his exact age is unknown. Shirtless, wearing shorts and sandals, he is a charismatic figure with flowing black hair and dark eyes that sparkle when he grins.

The first recorded contact with the Kayapo was just over a quarter century ago, in 1965, and since about 1977 their culture and way of life have been under siege. This year, during the 500th anniversary of Columbus's first contact with the Americas, Paiakan's village faces irrevocable change. Yet he greets the visitor warmly, speaking in Portuguese: "Your flight was safe. You are here. Everything is good."

After nightfall, as sounds of the forest fill the darkness, a Kayapo elder points to the stars and observes that they are distant campfires. Paiakan rests in a hammock under the roof outside his house speaking softly: "Since the beginning of the world, we Indians began to love the forest and the land. Because of this, we have learned to preserve it. We are trying to protect our lands, our traditions, our knowledge. We

Paiakan, chief of the Kayapo, meets with Anita Roddick of The Body Shop. The U.K.-based company purchases raw products from the rain forest, which are then used to manufacture cosmetics. With an alternative source of income, the people of the rain forest feel less pressure to sell their land to destructive loggers. (Courtesy of The Body Shop © Thomas L. Kelly)

defend to not destroy. If there was no forest, there would be no Indians."

On the Run

Rain forests are vital for the rest of us as well. Because they absorb carbon dioxide and emit oxygen into the atmosphere, rain forest destruction affects weather patterns and contributes to global warming, the so-called greenhouse effect. Furthermore, half of the world's biogenetic diversity is within these tropical forests, yet 50 percent of those species are still unknown to the outside world. About one-third of the world's medicines currently are derived from tropical plants, but indigenous people like the Kayapo have even more knowledge of plants with curative powers—knowledge that is quickly vanishing along with the forests themselves.

Among the Kayapo, preparation for a new leader begins at birth. Such was the case for Paiakan, who is descended from a long line of chiefs. His father, Chiciri, who lives in Aukre, is a highly regarded peacemaker; and when Paiakan was born, the tribe received a "vision" of his special destiny.

"When I was still a boy," Paiakan recalls, "I knew that one day I would go out into the world to learn what was coming to us."

As a teenager, Paiakan got his chance.

He was sent to the Kayapo village of Goro-tire for missionary school, where he met white men who were building the Trans-Amazonian Highway through the jungle. Paiakan was recruited to go out ahead of the road's progress, to approach the previously uncontacted tribes.

When he went back to see what was coming on the road, however, he saw an invasion of ranchers, miners and loggers using fires and chainsaws. As he watched them tearing down vast tracts of forest and polluting the rivers with mercury, he realized that his actual job was to "pacify" other Indians into accepting it.

"I stopped working for the white man," he says, "and went back to my village. I told my people, 'They are cutting down the trees with big machines. They are killing the land and spoiling the river. They are great animals bringing great problems for us.' I told them we must leave, to get away from the threats."

They are cutting down the trees with big machines. They are killing the land and spoiling the river. They are great animals bringing great problems for us.

Most of the Kayapo villagers did not believe him, arguing that the forest was indestructible. So Paiakan formed a splinter group of about 150 men, women and children who agreed to move farther away. For the next two years, advance parties went ahead to plant crops and build homes. In 1983, they traveled four days together, 180 miles downriver, and settled in Aukre.

"Our life is better here," Paiakan says, "because this place is very rich in fish and game, with good soil. Our real name is Mebengokre—'people of the water's source.' The river is life for the plants and animals, as well as for the Indian."

But the new security did not last. During the 1980s, most other Kayapo villages in the Amazon were severely affected by the relentless invasions. Along with polluted air and water came outbreaks of new diseases, requiring modern medicines for treatment. Aukre was still safe, but smoke from burning forests already could be seen and smelled. Paiakan, realizing that he could not run forever, made a courageous decision. He would leave his people again—this time, to go fight for them.

He went to Belem, the state capital, where he learned to live, dress and act like a white man. He learned to speak Portuguese, in order to communicate with government officials. He even taught himself to use a video camera, to document the destruction of the forest—so his people could see it for themselves and so the Kayapo children would know about it.

Fighting Back

Paiakan continued to travel between Aukre and the modern world, at one point becoming a government adviser on indigenous affairs for the Amazon. In 1988, when the rubber tapper Chico Mendes was shot dead by ranchers for organizing grassroots resistance to deforestation, it was feared that Paiakan himself might be a target.

"Many indigenous leaders have been killed," says Darrell Posey, an ethnobiologist from Kentucky who has worked with the Indians of Brazil for 15 years, "but publicity surrounding the Mendes murder may have helped to protect Paiakan." The Brazilian Pastoral Commission for Land has counted more than 1,200 murders of activist peasants, union leaders, priests and lawyers in the past decade.

In 1988, after speaking out against a proposed hydroelectric dam in the rain forest, both Paiakan and Posey were charged with breaking a Brazilian law against "foreigners" criticizing the government. Because Indians are not legally citizens, Paiakan faced three years in prison and expulsion from the country; but when other Kayapo learned of his plight, some 400 leaders emerged from the forest in warpaint. The charges were dropped.

"In the old days," Paiakan told the press, "my people were great warriors. We were afraid of nothing. We are still not afraid of anything. But now, instead of war clubs, we are using words. And I had to come out, to tell you that by destroying our environment, you're destroying your own. If I didn't come out, you wouldn't know what you're doing."

In 1989, Paiakan organized an historic gathering in Altamira, Brazil, that brought together Indians and members of the environmental movement. A major theme of the conference was that protecting natural resources involves using the traditional knowledge of indigenous peoples. "If you want to save the rain forest," he said, "you have to take into account the people who live there."

With increasing support, Paiakan acquired a small plane for flying to and from his village. He also made trips to the United States, Europe and Japan, even touring briefly with the rock star Sting, to make speeches about the growing urgency of his people's plight.

But the erosion of Indian culture in the Amazon forest was becoming pervasive. With the influx of goods ranging from medicines to flashlights to radios to refrigerators to hunting gear, village after village was succumbing to internal pressure for money to buy more. By 1990, only Aukre and one other Kayapo community had refused to sell their tree-cutting rights to the loggers, whose tactics included seductive offers of material goods to Indian leaders.

In June that year, racing against time, Paiakan completed negotiations for Aukre to make its own money while preserving the forest. Working with The Body Shop, an organic-cosmetics chain based in Britain, he arranged for villagers to harvest Brazil nuts and then create a natural oil to be used in hair conditioners. It would be their first product for export.

Hard Economics

Paiakan returned with his triumphant news only to learn that other leaders of Aukre—during the previous month, in his absence—had sold the village's timber rights for two years. It was a crushing blow, causing him to exclaim that all his "talking to the world" had been in vain. He said that if he could not save his people, he would rather not live.

"He went through a period of intense, deep pain," says Saulo Petian, a Brazilian from Sao Paulo employed by The Body Shop to work with the Kayapo. "He left the village and went far along the river, to be by himself. After about two months, when he got over his sadness and resentment, he came back and told me, 'Well, I traveled around the world and seemed to be successful, but the concrete results for the village were very little. These are my people. They have many needs. I can't go against them now.' "

So Paiakan made peace with the other leaders of his village and started over. "I was like a man running along but who got tired and stopped to rest," Paiakan recalls. "Then I came back, to continue my fight into the future."

What began was the simultaneous unfolding of two events, by opposing forces, in Paiakan's village. One was the beginning of construction by the Indians of a small "factory" with a palm-leaf roof for creation of the hair-conditioning oil. For Paiakan it was a way of showing his people how to earn money from the forest without allowing it to be destroyed. Meanwhile, loggers came through the forest constructing a road that skirted the edge of the village. By 1991, trucks were arriving from the frontier to carry back loads of freshly cut timber.

The white men left behind the first outbreak of malaria that Aukre had seen, mainly afflicting the elders and children. The only consolation for Paiakan was that the tree-cutters had just a couple of dry months each year when the road was passable.

"Through the Brazil-nut oil project," Petian says, "Paiakan is showing his people another possibility for satisfying their economic needs. He's giving them a viable alternative that includes helping to save the forest and their way of life."

Throughout Brazil, there is similar effort by environmentalists and Indian groups to discourage deforestation by creating markets for nuts, roots, fruits, oils, pigments and essences that can be regularly harvested. Since 1990, about a dozen products using ingredients from the Brazilian Amazon have entered the American market. The nuts, for example, are being used to produce a brittle candy called Rainforest Crunch. The candy is also used by Ben & Jerry's Homemade Inc. for one of its ice-cream flavors.

Paiakan is showing his people another possibility for satisfying their economic needs. He's giving them a viable alternative that includes helping to save the forest and their way of life.

"Paiakan is one of the most important leaders looking at alternatives for sustainable development," says Stephan Schwartzman, a rain forest expert at the Environmental Defense Fund in Washington, D.C. He cautions, however, that "nothing in the short term can compete economically" with cash from the sale of rights for logging and gold mining.

The Vision Spreads

Up to 8 percent of the two-million-square-mile Amazon rain forest in Brazil—an area about the size of California—already has been deforested. Once the trees are gone, the topsoil is quickly and irreversibly eroded, so that in just a few years hardly anything can grow, and both cattle-raising and agriculture become nearly impossible.

A hopeful sign is that Brazil's president, Fernando Collor de Mello, who took office in 1990, has taken some positive steps to protect both the forest and the Indians. (The population of indigenous people in Brazil, once at least three million, has fallen in this century to 225,000.) Last November, President Collor moved to reserve more than 36,000 square miles of Amazon rain forest as a homeland for an estimated 9,000 Yanomami Indians in Brazil. He also approved 71 other reserves covering 42,471 square miles, some 19,000 of which will be set aside for the Kayapo—about 4,000 people in a dozen villages.

It was a major victory for Paiakan, giving him more concrete evidence to show that his previous efforts outside the village had been worthwhile.

"Paiakan has a vision," Darrell Posey says. "He's trying in a lot of ways to maintain his traditions—setting up a village school for Kayapo culture, creating a scientific reserve. At the same time, he's making the transition to a modern world in which white men are not going to go away. He knows you either deal with them or you don't survive."

These days, Paiakan is working to organize an Earth Parliament of indigenous leaders in Rio de Janeiro in June [1992]. The global parliament will run simultaneously with the United Nations Conference on Environment and Development, the so-called Earth Summit, which more than 70 percent of the world's heads of state are expected to attend to ponder the fate of the planet.

"Paiakan has been at the center of incredible change, whether he has wanted to be or not," Posey says, "and now he's trying to straddle both the past and the future. I would hope that people in positions of power will see him as someone who can help the world turn back to its roots, to those whose lives depend on working with nature and not against it."

The Brazilian rain forest has taken on tremendous symbolic value worldwide.

The Brazilian rain forest itself has taken on tremendous symbolic value worldwide, says Thomas Lovejoy, a leading Amazon researcher and assistant secretary for external affairs of the Smithsonian Institution. "It's a metaphor for the entire global crisis," Lovejoy adds. "If we can't deal with that environment and with the people who live there properly, it's unlikely that we'll be able to deal with the rest."

At sunrise in the Kayapo village of Aukre, the red clay of the logging road is wet from rain. The trucks are gone now, and there is serenity as the tropical heat moves in. A shaman, or medicine man, is treating Paiakan's wife for an illness, using plants from the "pharmacy" of the forest. Some of the men are going off on a hunting trip. Women and children bathe in the river as butterflies of brilliant colors swirl across a blue sky. Time seems to stand still, before it races on.

Racing to Save Hot Spots of Life

By Michael Lipske

In coming to terms with the bird-finding skills of Theodore A. Parker III, it may help to imagine him as something more than a mere mortal man of science. Rather, consider picturing him as a dazzling example of high-tech ornithological hardware.

Lowered into the tropical rain forest by helicopter, the bearded, remarkably human-looking apparatus would be wrestled into position among shrubs and vines beneath the treetops. As the dawn chorus of bird song gathered strength, technicians in robin's-egg blue uniforms would flip switches on consoles wired to Parker. A few seconds of warm-up and whirring sounds would begin whispering from Parker's innards. Lights would flicker on a monitor. Shortly thereafter, the scientific names of bird species—*Amazona ochrocephala, Harpagus bidentatus, Tigrisoma lineatum*—would start spelling out on the electronic screen by ones and twos, then by tens and twenties and, ultimately, by the hundreds.

It helps to imagine Theodore Parker this way because he is a top-gun birdman with the sensitive ears of a jungle bat and the bottomless memory of a polyglot. Senior Scientist with the U.S. organization Conservation International, Parker can identify solely by sound more than 3,000 species of birds in the New World tropics—an unmatched ability he has honed in 18 years of scientific fieldwork and over the course of a bird-watching career he says did not get under way "seriously" until he reached the age of eight. Forthright about his immense skill ("I can go into a place and in five days of good weather I can find 80 to 90 percent of all the birds there," he says), Parker likens it to learning a language. Traveling or at home, he often unwinds by listening to his collection of bird recordings.

Lately, the scientist has been putting his gift to work in heretofore unstudied slivers and patches of South American rain forest believed to harbor unusually rich communities of birds, mammals, plants and other species—biological treasure troves that are also under such intense assault from commercial development and from exploding human populations, they may disappear altogether in a few years. These "hot spots," as the forest fragments have been dubbed by conservationists, may harbor as much as a fifth of Earth's plant species and a far larger proportion of animal species, while covering a mere speck of the globe.

Taking Inventory

Parker and his colleagues—a three-man, one-woman ecosquad made up of what one professional admirer calls "field-biology super-

In Ecuador, birdman Theodore Parker records the whisper of the wild as part of his surveying work. The talented ornithologist can recognize thousands of tropical bird species by the sound of their calls. (Randall Hyman)

stars"—make rapid assessments of the jungle hot spots, figuratively picking them up and shaking out a representative sample of their living contents. Brought together by Conservation International, the team (collectively known as the Rapid Assessment Program, or RAP) gathers information intended to help environmentalists and officials in host countries and the United States channel rain forest–saving efforts and scarce dollars toward protecting some of the neediest habitats on Earth.

Russell A. Mittermeier, Conservation International's president, calls Parker "a classic RAP scientist—a walking computer." But it will take more than a handful of superscientists

like him to stop the hemorrhage of extinctions from the world's tropical forests.

A green belt of life girding the planet along both sides of the Equator, tropical forests cover a mere 6 percent of Earth's land surface. Yet they are "utterly cluttered with species," writes biologist John C. Kricher in *A Neotropical Companion.* The Amazon forest, for example, may be the home of 30,000 plant species, twice the number of plant types found throughout the United States. A square kilometer (0.4 square mile) of South American forest can provide a home for hundreds of bird species and thousands of insect types. Forty-three ant species were counted on one tree in the Peruvian Amazon—more kinds of ants than are found in the entire British Isles. Some 1,200 plant species were recorded in less than a square kilometer of forest at a biological station in Ecuador.

Unfortunately, vast tracts of these fecund forests are being destroyed, frequently before scientists have even begun to tabulate their contents. Clearing by impoverished farmers as well as commercial logging claimed 42 million acres of tropical forest during 1990, according to a United Nations report. At more than an acre a second, that rate of loss is 40 percent faster than the speed of forest destruction ten years ago.

Snapshots

Western Ecuador provides one of the best, and most depressing, examples of how quickly a wealth of forest biodiversity can fade. A North Carolina–sized region, the western part of Ecuador once provided habitat for some 6,300 different lowland plant species. But in the last three decades, a mix of overpopulation and land clearing to grow bananas and other commodities has stripped forest from the region. Now, only 4 percent of western Ecuador's original forest still stands.

According to Theodore Parker, an ecosystem such as Ecuador's coastal moist forest could vanish altogether in another two or three years. Twenty percent of plant species in Ecuador's western region are endemic—found nowhere else on Earth. "We're talking," says Parker, "about a hell of a lot of species that are threatened with extinction."

"All of the hot spots are last-ditch efforts," adds Russell Mittermeier. And the list of death-row forests is long.

As catalogued by British ecologist Norman Myers, who coined the hot spots concept,

these rich but threatened forests span the globe and represent potential holocaust sites of plant and animal extinctions. Of the 18 sites Myers has identified, the Brazilian coastal forest, portions of western Ecuador, plus a rich forested strip along the eastern side of Madagascar (with some 6,000 plant species) are considered the most imperiled hot spots. Ranked right behind them are 15 only slightly less threatened species troves in Colombia, Western Amazonia, the Ivory Coast, the Eastern Himalayas, Malaysia, the Philippines and other parts of the tropics and subtropics.

A square kilometer of South American forest can provide a home for hundreds of bird species and thousands of insect types.

All told, the 18 hot spots support close to 50,000 endemic plant species. Because tropical forest inventories suggest the existence of at least 20 animal species for every one plant, the hot spots may contain 1.25 million species of insects, reptiles, mammals, birds and other creatures. For the RAP squad of biological hotshots, Myers's roster provides a focus for investigations. "Maybe by doing the quick-and-dirty survey work we're doing," says RAP botanist Alwyn H. Gentry, senior curator at the Missouri Botanical Garden, "we can know enough" to make more informed conservation choices.

In the effort to know more, Conservation International's RAP team twice descended on Ecuador during 1991. Parker, Gentry, Louise Emmons (a mammalogist with the Smithsonian Institution) and Robin Foster (a plant ecologist with the Smithsonian Tropical Research Institute) inventoried flora and fauna in various forest types along the range of coastal mountains that stretch between the cities of Guayaquil and Esmeraldas. At each site along the range, the biologists on the RAP team sought out a "snapshot" of Ecuadorian species richness.

*S*o far, only about 1.4 million species of plants and animals have been named by scientists, while some estimates put the total potential number at anywhere from 5 to 30 million.

Well before dawn each day, Theodore Parker entered the forest with his tape recorder, a unidirectional microphone and his educated ears. Traditionally, ornithological census work in the tropics has relied on mist nets, which work well for snagging birds in low vegetation. But, notes Parker, there may be another 150 or so species that only inhabit the forest canopy. With his ears alone, the scientist can cast a wider net, bagging whole communities of bird species that have escaped previous surveys and ferreting out isolated populations of poorly known species that are threatened with extinction.

While Parker listened for signature bird songs, Gentry set up one-quarter-acre transects to sample botanical specimens. Other team members stalked the forest in search of mammals and other wildlife.

As they worked, the scientists clearly could sense that they were in a neck-and-neck race with forest spoilers. Their entry into one "wilderness" area was via a logging road that was in the process of being cleared. At nighttime, trees loosened earlier that day by bulldozers would suddenly come crashing down. "The forest was literally disappearing before our eyes," says Gentry.

Spasm Control

So widespread is such habitat loss—particularly in tropical forests—that some researchers like Norman Myers say a major "extinction spasm" of plants and animals is under way, one that may wipe out millions of species around the world. But not all scientists accept that prognosis.

Doubters say it is impossible to predict future extinction rate when we still lack an accurate head count for the number of species on Earth. So far, only about 1.4 million species of plants and animals have been named by scientists, while some estimates put the total potential number at anywhere from 5 to 30 million.

Yet even without knowing the full measure of potential biocatastrophe, critics would be hard pressed to dispute Myers's contention that scarce conservation resources should be concentrated on areas that appear most vulnerable and that are believed to be profoundly rich in life forms. Save the hot spots, says Myers, and conservation planners will "get a bigger bang for each scarce buck invested."

"A lot of the wildlife efforts to protect endangered species have been rather unsystematic," says Myers. The ecologist notes that huge resources have been devoted to perpetuating an individual species rather than aimed at preserving an entire community of species, like those found in vanishing tropical forest hot spots.

On the other hand, does the hot spots/ rapid assessment approach really offer promise in helping to save forest species? Or will it only tell us in excruciating detail what surely is being lost?

Advocates of the approach point to results of a rapid assessment of Bolivia's Alto Madidi, a rain forest along an Amazon River tributary 200 miles north of La Paz. The RAP team's May 1990 visit there revealed that the forest is one of the ten richest sites for bird diversity in the world. Theodore Parker's dawn and dusk forays aurally netted 403 species—9 of them previously unknown in Bolivia. Team members also identified 45 species of mammals, including the exceedingly rare short-eared dog, a species only seldom seen in the wild by scientists. Gentry's transect yielded 204 plant species.

Such information, say RAP proponents, is helping conservationists and Bolivian officials press for assistance to protect the jungle jewel. According to a World Bank official in Washington, D.C., Alto Madidi is "one of the areas being explored for protection" under the bank's new conservation assistance program.

Bolivia, however, is blessed with relatively unsettled forests and a small population. That is not the case in another focus of RAP surveying efforts: Ecuador, where in the western part of the country a fragmented forest is under increasing pressure from poor colonists. Ecuador's population has more than doubled since the late 1950s, and may double again by the year 2025. In a paper published last year [1991], botanist Alwyn Gentry concluded, "Unless adequate protection is given to the remaining forest fragments . . . a major extinction of perhaps 1,260 endemic plant species can

be expected in western Ecuador in the very near future."

With such high stakes, conservationists argue that they need as much information as they can get to guide protection efforts. "There are millions of dollars out there for conservation projects, but millions of dollars won't go far," says Parker. When it is time to decide which sites out of many merit preservation, he adds, "The criteria should be biological, and the only way you can do that is to have groups like ours visit areas that in many cases have never been visited."

New Strategies

Over the coming year, the RAP scientists are expected to conduct high-speed surveys of forests in other South American countries, as well as in Africa and Southeast Asia. Even so, notes Parker, "The world is so big, we could have a thousand people doing what we're doing and we'd still overlook things." Fortunately, there is more than one way to save a forest.

Far from what one observer calls RAP's "sexy approach" to conservation, other people are rummaging through tool boxes for ways to patch up and preserve vanishing ecosystems. Often, the work is being conducted by ecocrats testifying in paneled hearing rooms or sifting through documents in windowless cubicles.

Four years ago, for instance, Conservation International began the implementation of debt-for-nature swaps. In these agreements, an environmental organization buys up a portion of a developing country's international debt; in return, the country's government agrees to put in effect local conservation programs or to preserve wilderness. The agreements also provide some relief from the country's share of an enormous Third World headache: the more than one trillion dollars the world's poorer nations owe banks and development agencies. Servicing that debt has led cash-starved governments to cut funds for environmental programs, and encouraged them to plunder resources.

The National Wildlife Federation has been instrumental in encouraging attention to environmental needs when loans are made by multilateral development banks (like the World Bank and the Inter-American Development Bank). Those development loans channel millions of dollars into poor countries for road projects, dam building, creation of plantations for agricultural exports and other useful efforts. But poorly planned projects funded by multilateral lenders have resulted in human colonization of rain forests and in agricultural projects that drained valuable wetlands that were habitat for wildlife.

The federation helped persuade Congress to hold hearings a few years ago that focused on the environmental impact of multilateral bank-financed projects, leading to some reforms in lending policies. "We're trying to change the economic development models followed in developing countries, to allow for better management of resources and protection of the environment," says Stewart Hudson of National Wildlife Federation's international programs division.

As a result of such efforts, global institutions are beginning to change the way they do business. The World Bank, for example, has established a $1.5 billion trust fund to attack loss of biodiversity as well as problems like global warming. And the World Resources Institute reports that money spent by U.S. private foundations on projects to preserve biodiversity increased 750 percent—to a total of $21.4 million—between 1987 and 1989.

Meanwhile, in another program arranged by Conservation International, rural residents in economically hard-pressed northwestern Ecuador are harvesting the ivorylike nuts of the *tagua* palm, which are used to make buttons on clothing manufactured by two U.S. firms. The idea is to create an economic incentive for keeping trees standing rather than logged in one of the hottest of the hot spots. Even on a small scale, such sustainable development projects offer great potential for helping both forests and people.

Hope Remains

Scientists like to point out that the greatest reason to preserve tropical forest diversity is its unknown value in yielding future medicines and other products. For proof, they point to the hundreds of food crop species that originated in the tropics and to drugs such as Vincristine, a rosy periwinkle extract that has proved useful in treatment of Hodgkin's disease.

As conservationists search for ways to ease pressure on tropical wild places and wildlife, one thing has become painfully clear. In the face of growing human populations, "parks and preserves," says Norman Myers, are at best "a partial answer."

"Setting up a protected area is a bit like building up an island in the face of the incoming tide," says the inventor of the hot spots concept. "And today the tide is coming in faster than ever. We have to find ways to deflect it."

Theodore Parker, whose amazing ears net birds during rapid surveys of jungle hot spots, believes that despite today's pressing population and poverty problems, there will come a time in Ecuador and other tropical countries when people will lament lost forestlands for reasons beyond simple economics. "They'll mourn the loss," says Parker, " purely for aesthetic reasons."

*S*etting up a protected area is a bit like building up an island in the face of the incoming tide. And today the tide is coming in faster than ever.

Parker has seen the future and finds it painful to look at. Unless some hot spots of rain forest are set aside quickly, he observes, "There's not going to be any forest left to protect" in places like western Ecuador. And that is why he continues his painstaking work beneath the rain forest canopy. "I'm a hopeful kind of guy," he says, "hopeful that we can still save a lot of important species."

From *National Wildlife,* April/May 1992. Reprinted by permission.

 EARTH CARE ACTION

Drawing the Line in a Vanishing Jungle

By David M. Schwartz

At dawn, the owl-like hoot of a rufous motmot penetrates the fog of Ecuador's coastal rain forest. All around me, drops from last night's downpour cascade from the leafy canopy more than a hundred feet overhead. Shrouded in the eerie mist, I am completely disoriented, but fortunately my Awà Indian companions know the way.

We slog through calf-deep mud, pulling hard to extract our boots at every step. The going is tough but worth the effort, I remind myself, for soon we will arrive at a sight unique in all the jungle—the *manga*. To the casual observer, the manga is a simple, broad path snaking through the underbrush. To the Awà, it may be the fine line between survival and ruin.

The Awà Indians that live along the border of Ecuador and Colombia are among the many native populations sprinkled across the map of South America. About 2,200 Awà make their home in Ecuador, and another 5,000 in Colombia. For centuries, these people have led lives of quiet isolation, virtually unknown to the outside world until about a decade ago.

The Awà are part of a complex ecological web of dense flora and teeming fauna in what may be the most biologically diverse quarter-million acres in the world. By taking only what they need, the Indians have preserved the 300 species of plants and animals they harvest for shelter, food and fiber. They seem the model of a people at one with their environment. But that environment is critically endangered—and so, as a result, are the Awà.

My trek to the heart of Awà territory has taken me to a remote village in northwestern Ecuador, a journey that required a full day's paddling in a dugout canoe. Welcomed as a guest of the Awà, I am getting a firsthand look at a people struggling against overwhelming odds to save their place in the rain forest.

Forces of change bear down hard on the region, and the Awà are caught in the squeeze. To the east, cattle ranchers have carved rangeland out of vast tracts of forest. In the west, loggers are stripping the land. To the north, in Colombia, plantations have transformed the jungle into immense monocultures of palm trees grown for oil. From all sides, *colonos* (colonists) have invaded Awà lands, clearing trees, overhunting game, fishing with dynamite and settling wherever they like.

*A*t the current rate of *deforesetation, the Pacific coastal rain forest of Ecuador will be completely gone by the year 2000.*

"At the current rate of deforestation," says Carlos Villareal, a government economist who works with the Awà, "the Pacific coastal rain forest of Ecuador will be completely gone by the year 2000, and the acute social and economic problems that already exist will blow up."

About six years ago, the Awà began an extraordinary battle against the forces working to destroy their home and way of life. With the help of a former U.S. Peace Corps volunteer and a national coalition of Indians, the tribe established its first government, the Awà Federation, and built shelters for community centers. Within just a few years, these hitherto uneducated tribespeople had built schools and convinced the Ecuadorean government to train members of their community as teachers and health-care educators.

The Bulldozers Stop Here

At times the Awà have had to reinforce their claims to the land with steel and gunpowder. More than once, they have taken up machetes and primitive muzzle-loading *escopetas* to scare away intruders. But no weapon in the Awà arsenal has proved as powerful as the novel land-management technique they call the manga.

The manga is nothing more than a 150-mile-long serpentine swath encircling Awà territory, carved out of the very rain forest the Indians wish to preserve. It is also nothing less than a demonstration of the power of positive action, and a model for other indigenous peoples trying to discourage deforestation on their own turf. In 1988, largely because of the manga, Ecuador's government designated Awà lands an "Ethnic Forest Reserve," South America's first such sanctuary. The Awà Reserve represents a stunning victory not only for the Indians, but for all the species that share their threatened home.

As my Awà guides and I make our way through the forest toward the edge of the reserve, the mist lifts. Hazy sunlight dances off the Rio Bagata. After fording the river, we reenter the jungle, darkened by the filter of leaves, lianas and tree trunks. We trudge uphill for 15 minutes. Then the light intensifies once again as we emerge in an open corridor: the manga. At first sight it is unremarkable—a clearing dotted with cultivated trees. But as my companions proudly explain, stripping this land was the best thing they ever did to preserve the forest.

Whisking their machetes, the Awà go to work clearing out the tangle of weeds that have invaded since their last visit. In just a few months, second-growth trees such as balsa have sent up shoots eight inches thick and 30 feet high. It takes a trained eye to distinguish unwanted plants from the cash crops: *borojo*, whose mineral-rich fruits can fetch more than a dollar apiece in Colombia; *qualte*, a sturdy wood used for home construction; *bacao*, a wild chocolate; *chonta duro*, a spiny-trunked palm that produces tasty pulp; and familiar crops— breadfruit, soursop, cacao, coffee—which can be sold for a profit.

Enclosing a Domain

It will be years before any of the plants can be harvested, I'm told. Sensing my disappointment, a young Awà boy wielding a machete as long as his arm chops down a small palm and deftly removes the heart. I accept it as a delicious snack.

To a people whose jungle is their castle, the manga is the "moat" that keeps intruders at bay, though its surface is solid ground, essentially unguarded and easily crossed. "The manga announces our land and tells everyone that the Awà live here," says Julian Cantincuz, president of the Awà Federation. In the borderless expanse of jungle, he says, colonos who

know exactly where the Awà lands begin will be less likely to invade—especially when the boundary is trumpeted by a clearing that, like a line drawn in the sand, proclaims, "Cross this and you're in for a fight."

But more than a tool for self-defense, the manga is also a rallying point for community organization. "The manga assures that we will work together," says Cantincuz. "So as long as the Awà are united, no one will take advantage of us."

Standing less than five feet tall, Cantincuz, like many Awà tribesmen, has a physical strength that belies his stature. Some of that strength is no doubt the result of the many hours he spends with his fellow tribesmen swinging a machete and planting crops in the manga.

Indeed, the sheer physical labor the manga represents adds immeasurably to its value. A project so large and tangible commands respect from Ecuador's government, which is often skeptical of Indian land claims. "When we talk to the government officials, the manga backs us up," says Cantincuz. "They know this is Awà land. We have the manga to prove it."

In Ecuador, as in most of South America, indigenous peoples cannot control the land they have occupied for millennia unless they obtain legal title. The government, citing the need for economic dvelopment, prefers to award title only to those who promise to "improve" the forest by putting it to "productive" use. Usually that means cutting it down. Time and again, poor and politically powerless Indians have fallen victim to nonindigenous populations who get the nod from the government to exploit the land for their own short-term goals.

The Awà took their first steps toward self-determination in the mid-1980s, when they learned that an agricultural cooperative had designs on a 6,000-acre plot in Awà territory. Meanwhile, a number of timber companies were planning to purchase wood rights nearby. Some had already intimidated Awà tribesmen into putting their thumbprints on contracts selling all their land for less than one cent an acre.

Tribal leaders called an emergency meeting and invited a man named Jim Levy to attend. A former Peace Corps volunteer who had helped the Awà organize to defend their territory, Levy put forth an idea: Why not cut a giant trail around all the tribe's land in one continuous swath? Not just a narrow footpath easily ignored and quickly overcome by jungle growth, but a 30- to 45-foot-wide manga (line of demarcation) enclosing the tribe's domain.

The Awà saw that the manga would help identify their land. Levy pointed out that it could do much more. With planting and cultivation, fruit and nut trees might grow in the manga and someday bring needed income. With a little extra vigilance, tribesmen posted to keep the manga clear of jungle overgrowth could serve as sentries, watching out for intruders. *Colonos* who crossed onto Awà lands could be ejected at once.

Most important, cutting the manga meant, in the eyes of the government, that the Awà were putting their land to "productive" use. The manga was the tribe's best hope for establishing a claim to ownership, possibly even title.

"The most striking thing the Awà have done with the manga is demonstrate to themselves and the country that you don't have to take an environmentally destructive and ecologically shortsighted approach to prove that you are using land," says Ted Macdonald, project director for Cultural Survival, a Massachu-

setts-based group that helps indigenous people like the Awà gain economic self-sufficiency.

In 1986, evidently hearing the message. Ecuador's government responded by mapping out the manga and awarding long-denied citizenship papers to the Awà. Two years later, the government created the Awà Reserve. While less secure than title, reserve status prohibits logging and keeps intruders out, thus enabling the Awà to manage their own natural resources.

Pressure Continues

The system has proven so effective that others want to try it as well. Across the border, the Awà of Colombia face most of the same problems as their Ecuadorean brethren. Having seen the manga work for their kin, the Colombians now plan to create one of their own.

At least two other tribes, the Huaorani of eastern Ecuador and the Embera of Colombia, have already started cutting mangas around their land. Meanwhile, the directors of the Rio Palanche Reserve, Colombia's biggest wildlife sanctuary, are considering using the manga concept as a land-management tool.

The creation of the Awà Reserve has heartened not only indigenous peoples but biologists and conservationists worldwide. With 300 to 400 inches of rainfall a year, the reserve is one of the rainiest places on Earth. Because rainfall correlates closely with species numbers, it is also wonderfully diverse. "Biologically, this is probably the richest area in Ecuador," says Calaway Dodson, senior curator of the Missouri Botanical Garden and director of Ecuador's National Herbarium. "And Ecuador is biologically the richest country in the world," he continues. "It is definitely an area worth saving."

Dodson and other biologists who have

studied the area credit it with uncounted thousands of plant species, many rare fish and amphibians and more than 600 birds, along with 5 cats, 4 monkeys and the highly endangered spectacled bear. About 40 bird species and 6 percent of the plants in the reserve are believed to be found nowhere else.

Conservationists are beginning to recognize that the key to saving the rain forest is to join forces with the people who live there. With that goal in mind, environmentalists from around the world met in Peru last year [1990] with representatives for 327 South American tribes and pledged to work together. Even so, optimism and international support are no assurance that indigenous people like the Awà will succeed in protecting their land. Too many hazards lurk on the murky edges of the political and economic landscape.

An island of protected habitat in a sea of environmental degradation, the Awà Reserve may ultimately be swept aside. Communal land title remains an elusive goal for the Awà, and without it the government could dissolve the reserve with a stroke of the pen. Furthermore, the "gold fever" that has infected much of South America could easily spread inside the reserve. Gold-mining companies already operate just outside its borders, and because the reserve's protected status doesn't exclude mining, there's little to stop them from moving in.

The most pressing danger to the Awà, however, comes from their 70,000 neighbors in the San Lorenzo area, the country's most impoverished corner. These nonindigenous people—nearly half of whom live by wood exploitation—are as bent on consuming the rain forest as the Indians are determined to sustain it.

After leaving the manga, three of my Awà companions accompany me to a community house where we will spend the night.

Along the way, a young man stoops to point out a log set between two rows of wooden pegs—a deadfall trap baited for *raton puyazo* (spiny rat), an important meat source. "Each family has at least 50 traps spread out over a large area," the man explains. "Once we make a catch, we do not use that trap again for three months. This way there will be puyazo for our children."

He disappears down the trail while I linger at the trap. This, I realize, is the fundamental difference between the Awà and their neighbors: The Indians are concerned about leaving natural resources for their children. Outsiders concentrate only on what they can take today.

"The reserve is one of the last stocks of intact natural resources in the region," says Carlos Villareal. "When the rest is gone, the people of the area are not going to sit back and starve to death quietly." Since 1983, Villareal has directed a small government agency called the Unidad Técnica del Plan Awà (UTEPA), which acts as official advocate for the Awà. Since the creation of the Awà Reserve, UTEPA has focused on helping the area's other nonindigenous populations find economic alternatives to plundering the rain forest and selling the spoils.

Projects funded by UTEPA so far include demonstration farms, where pigs are raised on sugar concentrates to show *campesinos* they can earn as much by raising swine on 2 acres of sugarcane as by ranching cattle on 20 acres of pasture. At one site, pig manure feeds fish in artificial ponds, and sludge from the ponds is used to fertilize sugarcane.

Saving the Chain

What began as an obscure Indian tribe's struggle **to protect** its homeland has evolved into a complex network of interrelated projects. And what appeared a few years ago as a dark future for the Awà and their neighbors is now brighter—still uncertain but streaked with hope.

After leaving the Awà, I stop in San Lorenzo, a raw-edged city of poverty and disease, before beginning my journey home. I depart at dawn on the *autoferro,* a converted bus mounted on a railway chassis that hairpins up the west slope of the Andes, affording a panorama of the Awà Reserve. Its great green vastness falls away to the horizon, so tranquil yet so tenuous in its coveted lushness.

I ponder the many interrelated pieces of this huge puzzle and wonder what lies ahead for the region. I recall something Carlos Vil-

lareal told me over lunch on the day I left for the reserve. "Just as you can't talk about protecting birds and forget about the trees," he said, "you can't talk about protecting the rain forest and forget the people who live there." On a napkin he drew interlinking ovals to represent the interdependent elements of the ecosystem.

"And if you talk about protecting indigenous people," he continued, "you can't forget the nonindigenous people who live all around them." Now, with all the links in place, there is finally a chance to save the chain.

From *International Wildlife,* July/August 1991. Reprinted by permission.

EARTH CARE ACTION

A Student's Crusade

By Nancy Phillips

As a college freshman two years ago, Find Findsen began researching the destruction of the world's rain forests, where thousands of acres filled with exotic plants and animals were being lost to development each day.

Right there in his Princeton University dorm, Findsen decided to buy some rain forest.

That impulse grew into the Rainforest Conservancy, a student group dedicated to preserving tropical forests and the plants and animals they nurture. In little more than two years, the crusade has spread to 45 campuses, won praise from environmental groups and attracted both corporate contributions and a star-studded board of directors.

But Findsen emphasizes that it wasn't easy. He ran up a $2,000 phone bill the first month, calling environmental groups around the globe for advice on how and where to buy land.

Then he founded the conservancy.

That first year, the group raised more than $5,000 to buy 142 acres of rain forest in northwestern Belize. There, for $37 an acre, the students bought part of a tropical forest in a preserve run by a coalition of environmental groups.

Last year [1990], Findsen's group, which he described as "a dorm-room operation,"expanded to 44 more campuses across the country and raised more than $15,000 to buy 333 acres of forest to add to its preserve in Belize. This year, it hopes to add more.

"I think each age has a challenge, and I think this is ours," said Findsen, 23, a native of Denmark who is majoring in political science. "I really see it as a personal imperative. I think this problem is crucial, and we have to act."

Sitting in the university's student center, Findsen ticked off facts about rain forests like a biology professor. They are home to more than half of the world's plant and animal species and are being destroyed at a rate of 75 acres each minute to make room for development, Findsen said. Twenty-five percent of all medicines are derived from plants found in tropical forests. More than 1,300 rain forest plants are being studied as possible cures for cancer. And scientists have identified only 1 percent of all rain forest plants, so there could be scores of potential crops and medicinal plants among the rare species in the forests.

"By destroying the rain forest, you have the scenario for a nightmare," said Findsen, an intense young man with sandy hair and blue eyes. "This is a problem that's just so critical."

Spreading the Word

His group takes that message to almost anyone who will listen—from fellow students to corporate executives.

"We started with our friends, trying to get them to donate money, but then we ran out of friends," he said. "So we started calling other schools, trying to get people involved."

As Findsen's crusade stretched to campuses across the country, the work grew increasingly time-consuming. Last year, he and his colleagues devoted hundreds of hours to the cause each week, searching for new members and new promises of charitable and corporate support.

Term papers and exam reviews were delayed until the last possible minute. Weekends were sacrificed. Even summer vacations. All for the good of forests far away.

"Sometimes I get sad about it, and I just want to quit," said Findsen, who learned the ups and downs of volunteer work when he helped raise money for the hungry as a teenager in Copenhagen. "But it's so important that we just keep working at it."

The work has its rewards. Although the Rainforest Conservancy is just two years old, it has gained recognition and praise from prominent environmental groups. Its advisory board of directors includes John C. Sawhill, president of the Nature Conservancy; Katherine Fuller, president of the World Wildlife Fund; former Environmental Protection Agency head William D. Ruckelshaus; and Thomas Lovejoy, environmental adviser to President Bush and an administrator at the Smithsonian Institution.

With their help, the conservancy stepped

up its fund-raising efforts this summer. Working in borrowed offices a few miles from campus, Findsen and his colleagues lined up promises of support from businesses and charitable organizations across the country.

---

*S*ometimes I get sad about it, and I just want to quit. But it's so important that we just keep working at it.

Chemical Bank donated the use of telephones at its Princeton branch a few evenings a week. A manufacturer of recycled paper agreed to share its profits with the conservancy in exchange for help in booking new orders. And students' contributions continued to pour in.

This school year, with fuller coffers, the conservancy plans a huge expansion. Findsen wants to start 200 new chapters, host a national conference on rain forests in the fall and buy as many more acres of rain forest as his budget will allow. His goal is to raise $500,000 this year. An ambitious target, but Findsen's

supporters have launched an all-out effort to pull if off.

"I have faith," said Donna Feinberg, a Swarthmore College student who spent the summer working for the conservancy. "We're tackling an enormous problem, but we've gotten really good response because people realize that there's a lot to be gained here."

When school begins next month [September 1991], the group will leave its borrowed offices, but the work will continue in dorm rooms equipped with computers and fax machines that donations helped buy. Findsen, who waited 18 months to be reimbursed for his $2,000 investment in that initial, mammoth phone bill, still shells out some of his own money from time to time.

But he said it's worth it. For the good of a tropical preserve filled with gray foxes, monkeys, pumas and jaguars, 400 species of birds and thousands of trees and flowers. "We're making a tangible contribution," Findsen said. "We want to preserve as much of the rain forest as humanly possible."

Reprinted with permission from *The Philadelphia Inquirer,* August 27, 1991.

Peru's Rain Forest: The Kindest Cut

By Kathryn Phillips

Patches of bare pastureland scar Peru's lush Palcazu Valley. A half-century ago, Swiss and German colonists introduced cattle to this isolated, rain-soaked region, stripping trees from virtually every riverbank to accommodate their herds. More recently, Peru's government planned to import 30,000 colonists into the area from around the country. The aim: to turn this valley east of the Andes into a vast logging and cattle-grazing region.

Like most residents, 37-year-old Emilio Sanchoma saw ranching as the only hope for his family's future. A dark-eyed Yanesha Indian whose ancestors arrived in the area 4,000 years ago, Sanchoma figured he, too, must destroy his forest home to survive.

Now it seems he figured wrong. Today, forest loss in Palcazu has not spread much farther than the riverbanks. The government has dropped its sweeping plans. And fellow tribesmen have elected Sanchoma *jefe de campo*, or chief of forestry, for an innovative new program in sustainable forestry management.

Using what's called a strip-shelterbelt system, which relies on the forest's natural ability to reseed itself, Sanchoma and fellow tribe members are reaping profits from local woodlands without wiping them out. The program is only part of a long-term conservation strategy, researchers caution. Yet results are so encouraging that it has become a model for other indigenous people locked in a cycle of poverty and destruction that claims 2.5 acres of the world's rain forest every second.

The idea came from a group of scientists who had come to Palcazu in 1981 to look into the potential impact of the government's massive development plan. These dozen or so anthropologists and foresters funded by the U.S. Agency for International Development saw the need for a program that would make the forest so valuable that ranching would lose its appeal.

---

*S*anchoma and fellow tribe members are reaping profits from local woodlands without wiping them out.

At first, the Yanesha were less than receptive; history had made them leery of outsiders and their promises. But after two years, 5 of the valley's 11 Yanesha communities formed a committee to look into the proposal. The result was the Yanesha Forestry Cooperative, which, under the leadership of elected tribe members, began a logging and sawmill operation in 1985 using the strip-shelterbelt system.

Mimicking Nature

The concept is based on research by tropical ecologist Gary Hartshorn, now vice president for science at the World Wildlife Fund. He discovered that many tropical trees grow only in natural gaps in the forest. People can mimic these gaps, he suggested, by cutting trees in narrow strips and leaving wide sections of forest intact. With the help of seed-carrying animals, the gaps regenerate in 30 years—a fraction of the time it takes a razed forest to replenish itself.

Today, more than 200 Indians earn their living as members of the Yanesha cooperative. Workers with chain saws cut strips 30 yards wide and 300 yards long through the forest. Oxen haul the logs to a processing plant, where other tribe members convert them into boards, fence posts and charcoal. From an airplane, the harvested strips are barely visible. "You would be hard-pressed to find them," says Robert Simeone, a forestry consultant who serves as an adviser to the Yanesha.

The Indians have found a developing market around the world for their ecologically benign wood products. Project workers also act as traveling consultants, sharing their expertise with tribes such as the Quichua of Ecuador's Napo Valley, who recently launched a similar program. Now, with luck, perhaps Emilio Sanchoma and others struggling to save their homes can finally see the forest for the trees.

From *International Wildlife*, May/June 1992. Reprinted by permission.

Village Committees Guard Endangered Forest in Bangladesh

By Barbara Crossette

In Pingabaho, Bangladesh, a grassroots environmental movement is growing, bringing awareness and strength to villagers struggling to subsist on the fringes of vanishing forests in this overcrowded, climatologically vulnerable South Asian country.

Radhakantha Burman is one of its local heroes. For two years Burman, a tribal village leader, has been guarding a 200-acre patch of sal trees, or *Shorea robusta*, a tropical hardwood. The trees stand on land that is in theory a common woods protected by forestry officials. But corruption, thuggery and unchecked erosion had combined to deplete the land. Wood for farm implements and leaves for mulch and cooking fires were getting harder to find.

Aided by surveillance squads from five hamlets, Burman has in two years been able to deter illegal loggers and encroaching landowners looking for more space to plant rice.

"We used to guard it around the clock," he said. "We only have to keep an eye on it now."

For battered Bangladesh, this is a success story. But it is also part of a trend in South Asia, where individuals and public action groups, backed by international foundations and aid programs, are taking national development into their own hands, despairing of costly government programs that never seem to reach the poor.

A Wider Look

In Bangladesh, a Dhaka-based nongovernmental organization called Proshika has taken the lead in forestry protection. With help from the Ford Foundation, Proshika (its name is an acronym for, roughly, "training, education and work") is assisting village committees not only in setting up forestry patrols but also in developing and diversifying their small farms to improve general living standards.

"Local problems should be solved by local people, but the nongovernmental organizations are necessary because governments are not giving us what we need," said Subodh Chandra Sarkar, the chairman of Proshika's Kaliakor regional committee, in an interview in nearby Fulbaria, the area's largest town.

"These programs teach us to analyze and look at problems in a wider way," Sarkar said through an interpreter. "We all knew about the firewood crisis, but we weren't aware how this problem of ours related to erosion, the depletion of soil, the loss of plants and animals and so on."

Mafrusa Khan, an economist who is a project coordinator for Proshika in Dhaka, says that a number of factors had contributed to the depletion of the sal forests of central Bangladesh, one of three remaining forest areas in the country. The other two are in the mangrove areas of the Sunderbans and the Chittagong Hills.

The use of wood for fires, tools and village construction is only a small part of the problem because sal trees grow back quickly from stumps, at the rate of several feet a year. More dangerous in the long run, Khan said, are illegal logging operations, in which stumps and roots are also burned to make charcoal; the clearing of land for farm crops; and the substitution of quick-growing plantation trees like eucalyptus or monocrop bushes for the hardy sal tree.

Logging operators also cut roads into the soft soil, speeding erosion and ultimately desertification. Wildlife has no place to hide and breed.

A Forest Recovering

In Pingabaho, a tribal hamlet of mud houses populated by a mixture of aboriginal people and a few Bengalis, the villagers welcomed Proshika organizers, whose work they had heard about. Village leaders had gone to look at other Proshika projects in 1987 and 1988.

Two years ago [1989], people from Pingabaho and nearby hamlets joined a training program. They identified the forest they wanted to protect and divided it into zones for patrolling, Burman said. Local forestry officials, often as powerless as villagers when politically powerful vested interests are involved, have been generally sympathetic.

The vigil has been nonviolent in Pingabaho, more than ten miles from the nearest motorable road. In other areas, grassroots forest protectors have been assaulted by rich landholders staking claims to common land, much of it in litigation since the partition of British India in 1947 and the creation of Bangladesh in 1971.

Pingabaho's sal forest is now recovering, and there are plenty of sticks to be collected from the earth for trellises and frames to support new vegetable and spice crops. Spare underbrush is sold for firewood, always in great demand in Bangladesh, producing cash income for the occasional luxury.

"This year, we sold enough branches and twigs to pay for our children's football matches," Burman said.

Fighting for a Culture

By Sally-Jo Bowman

Three years ago [1989], Noa Emmett Aluli was among a group of native Hawaiians that marched barefoot for several miles on a new crushed-lava road to the last standing ohi'a tree in a clearing at a geothermal energy drill site. The clearing is located in the heart of the Big Island of Hawaii's Wao Kele o Puna—a 27,000-acre gem on the slopes of the world's most active volcano that is the last sizable lowland tropical rain forest in the United States.

Under the tree's scarlet blossoms, which are considered sacred to the Hawaiian fire goddess Pele, the Hawaiians built a rock altar. Within a week, both altar and tree were bulldozed down. For Aluli, it was one more example of the blatant disregard for native Hawaiian culture and religion, and the wanton destruction of island forest ecosystems.

The 48-year-old physician, who lives on the island of Moloka'i, has been actively fighting rain forest development in the fiftieth state for nearly a decade now. In 1983, he cofounded the Pele Defense Fund (PDF) to rally many of the state's 8,000 full-blooded and 180,000 part-Hawaiians to protect traditional religious and cultural sites. The group bases its activities on the Hawaiian concept of *aloha 'aina*, or "loving and caring for the environment of land, sea and air."

One of the main targets of its activities has been protecting the Wao Kele o Puna, where developers want to build a 25-megawatt

Noa Emmett Aluli doesn't take lightly the destruction of Hawaii's rain forests. (Minden Pictures © Frans Lanting)

geothermal electrical power plant to help the island become more energy self-sufficient. Aluli and other opponents contend that geothermal energy is not dependable or economical, and that it produces potentially dangerous hydrogen sulfide emissions. They also maintain that cutting down the rain forest is both a biological and cultural disaster.

To Aluli, the forest surrounding the volcano is sacred to Hawaiian religion. Drilling beneath it, he says, is the equivalent of jackhammering the Vatican floor to search for oil. "We also must remember," he says, "that an island ecosystem is degraded much faster than a continental one. Even small damage is irreparable." The soft-spoken physician argues that if the United States won't save its own rain forests, "what business does it have telling South Americans or Malaysians to stop destroying theirs."

Hawaiians for Hawaii

Two years ago [1990], in a show of support for the PDF, some 2,000 Hawaiians en-tered the fenced-in drill site in what is perhaps the biggest protest demonstration in the state's history. And last year, the Sierra Club Legal Defense Fund joined the fray when it obtained a U.S. court ruling requiring an environmental impact statement before more federal money for geothermal development in the islands can be released. The PDF joined in a similar suit to force the state to do its own impact study. Several months may pass before the controversy is finally settled.

Aluli's strategy, says Allan Kawada, an attorney who is representing the geothermal developer, "is to delay progress and eventually stop it." However, he concedes, he sees the physician as "the plaintiff, the opponent, but not the enemy. He's intelligent and articulate."

Aluli, meanwhile, is concerned about the future of his people. "Indigenous Hawaiians," he says, "are as endangered as the Hawaiian hawk of Wao Kele o Puna. But now there's a strong community of native people here that wants to leave a legacy for future Hawaiians."

From *National Wildlife,* October/November 1992. Reprinted by permission.

An urban ecology movement is alive and well, as evidenced by women working in a community garden in downtown Detroit. With efforts from volunteers such as these, numerous cities and suburbs are turning eco-green. (Jim West)

SMALL TOWN THINKING

By Erik Hagerman

Few American ideals have promised so much and delivered so little as the suburb. More than 40 years have passed since Levittown, New York, inaugurated the age of large-scale suburban housing developments, and what once epitomized the American dream is looking more like a nightmare.

Mile after mile of countryside has been overrun by faceless strip development festooned in neon and plastic. A once-liberating road and highway network has become a mire of congestion. Trimmed hedges and ornamental front lawns now create secluded refuges rather than extensions of community.

Finally, in its shortsighted attempts to bring people to nature, the suburb has ended up defiling the environment. There is no aspect of modern life that has generated more environmental damage than our car-dependent transportation system—from the alarming amounts of raw materials it consumes, to the pollutants it puts out, to the very roads themselves—and sprawling suburban development has played a primary role in shaping this system.

This may be about to change. A small but increasingly influential group of architects and town planners, known as "neo-traditionalists," are turning disappointment with suburbs into blueprints for a new kind of community. Their inspiration is none other than the American small town.

For this new generation of town-builders, the appeal of the small town lies not in nostalgia but in functionality. "The problem with current suburbs is not that they are ugly," says Andres Duany, who, with his wife, Elizabeth Plater-Zyberk, runs an architecture and planning firm based in Miami, Florida. "The problem is they don't work."

The Small Town Model

Small towns do work, largely because they were designed for people, not cars. According to Duany, "Most of the needs of daily life can be met within a three- to four-acre area, and generally within a five-minute walk of a person's home." Just about any town of less than 50,000 people built before the turn of the century can serve as an example, as can the older sections of cities such as Annapolis, Maryland, and Charleston, South Carolina.

The ill-fated love affair with the auto changed this time-honored design. As planners switched their focus from walking to driving, they unintentionally dismantled many of the basic physical features that made American communities such pleasant places to live. Roads were widened and parking lots expanded; sidewalks and tree-lined streets were eliminated.

> *The problem with current suburbs is not that they are ugly. The problem is they don't work.*

The neo-traditionalists want to return these features to the American townscape. In the process, they hope to bring people out of their cars and perhaps return some vitality to what they see as an ailing sense of community.

The cornerstone of the new approach is the concept of mixed use, which brings homes near offices, and both near shopping. Add to this low buildings, streets that encourage walking and downplay cars, and nearby parks and town squares, and you have the basic framework of a viable small town.

Real, operating examples of the new traditionalist approach are few because of

years of lag time between design and development. The furthest along of these projects are just beginning to open, but they are easy to distinguish from most of the suburban development that is now taking place.

At Kentlands, a neo-traditional community being built in Gaithersburg, Maryland, the sections of neighborhoods that have been built—with their narrow, gridlike streets, compactly sited houses and back alleys lined with garages—bring to mind not suburbia but nearby Annapolis, Maryland, and Georgetown in Washington, D.C. It will be ten years before the development fills out its design of a town of close-to-home neighborhood parks, corner stores, public lakes, elementary and preschools and a downtown center. But Kentlands already evokes the sense of place that most suburbs noticeably lack.

It's this ineffable quality that lies at the heart of neo-traditionalism's rapidly growing popularity. "[Neo-traditionalism] is unique in the history of modern architecture in that it has been as much a popular as a professional phenomenon," says Vincent Scully, art professor at Yale University and one of the country's most influential architecture critics. Duany and Plater-Zyberk, he says, are "bringing to fruition the most important contemporary movement in architecture."

Scully's comments are borne out by a recent Gallup poll. Asked where they would prefer to live, more people chose a small town than suburbs, farms or cities.

These desires have not been lost on developers. Many have gotten no further than a nostalgic image of neo-traditionalism and continue to peddle what are essentially the same old subdivisions, dressed up with a sidewalk here and a gazebo there. "Like a frog turning into a prince," Phillip Langdon recently wrote in *The Atlantic,* "the [conventional] pod becomes a 'village' with a kiss from the marketing staff."

But an increasing number of developers are attempting to build the real thing. Over the past eight years, Duany and Plater-Zyberk have designed more than 30 new towns and urban retrofits, from Birmingham, Alabama, to Los Angeles, and ranging from 60 acres and about 100 people to 10,000 acres and more than 25,000 people.

Peter Calthorpe, an architect and town planner based in San Francisco and another pioneer in the neo-traditional approach to community design, is finding himself in similar demand. His list of current projects includes Sutter Bay, one of California's largest residential development projects. The project will consist of 14 villages and house 175,000 people on a 25,000-acre site ten miles north of Sacramento. What makes Sutter Bay particularly unique is that Sacramento has already agreed to extend its new light-rail transit system to link the community with the city center.

Breaking Codes

Unfortunately, in most regions of the United States, actually building a small town might get you arrested. Many of the distinguishing characteristics of small towns—narrow streets, on-street parking, shops near residences—are forbidden by codes written when auto dependence was thought to be a sign of progress.

As a result, many neo-traditionalists are concentrating their attention on persuading local zoning boards to revise these codes. Says Duany: "Planning codes and zoning ordinances are the genetic codes that determine what communities will look like in the future. It's unrealistic to think that we can retrain 35,000 planners. The most effective route is to get this [kind of town design] in the codes, and then let them follow it."

Americans finally seem to be recognizing something Europeans have known for quite some time: People can guide and control the shape of their communities.

In each of their many projects, Duany and Plater-Zyberk helped communities push through revised sets of zoning codes to reallow small-town features. To spread the word further, they recently completed a set of neo-traditional ordinances that virtually any community can add directly to its existing zoning codes. Calthorpe has put together a similar package for the county of Sacramento, which asked him in 1990 to design a plan that would make transit and pedestrian orientation a part of all new development in the county.

Where the bureaucratic barriers to neo-traditional communities have come down, the public has responded. At Duany and Plater-Zyberk's first project, Seaside, a mixed-use resort community on the coast of the Florida panhandle, lot prices have increased 500 percent in the past eight years. Seaside is still the only fully completed example of the neo-traditional approach, and it is somewhat unrepresentative because of its resort status, but 12 other projects are now under construction, with homes at virtually all selling at or faster than the market pace.

Americans finally seem to be recognizing something Europeans have known for quite some time: People can guide and control the shape of their communities. The recognition comes none too soon. As Duany noted at the end of a recent lecture, "[Unless things change], all the energy that we put into all this growth is going to be the heritage of misery. If we're not careful, we are going to be remembered as the generation that destroyed America."

From *World Watch*, July/August 1991. Reprinted by permission.

The Greening of the Big Apple

By Will Nixon

"How can you live in Manhattan and call yourself an environmentalist?" a suburban friend once asked me. Before I could answer, she launched into her own New York nature story: how she had worked for a while at The Body Shop in Greenwich Village selling rain forest shampoo or Amazon bath oil or something like that, and she had come into the store early one morning to find a strong odor The Body Shop did not sell. It was putrid, in fact. It was, she discovered after searching high and low, a king-sized rat wedged behind the steam radiator, slowly cooking. She doused the poor creature with cruelty-free perfume until the superintendent came to take away the roasted corpse. But the rat wasn't even what drove her out to the suburbs. She just got tired of waking up every morning to a fresh dusting of black soot on the windowsill where tomatoes could have been ripening.

I had to sympathize. I knew I had lived in New York too long when I found myself going to all those Vietnam movies several years ago for the great vegetation shots. New York City is not the natural world. Our biggest blizzard of the year came on June 10 of this year [1991] when Wall Street's faithful dumped 87 tons of recyclable paper on the Operation Welcome Home parade. Our best crop is concrete. Our diesel buses chug along, emitting black clouds like mechanical squids. And our rivers turn into toilet bowls every time heavy rains cause our sewage system to overflow. No wonder, come weekends, any New Yorker with a really disposable income heads for the Hamptons of Long Island or the north woods. Vermont's former deputy commissioner of natural resources, Mollie Beattie, once lamented that her state had become "New York City's playground."

I knew I had lived in New York too long when I found myself going to all those Vietnam movies several years ago for the great vegetation shots.

Yet, if given half a chance, the city can surprise you. As the head of the Gaia Institute at Manhattan's Cathedral of St. John the Divine, Paul Mankiewicz is often out and about the city, preaching the wonders of ecology. He'd like to turn 10 percent of our rooftops into greenhouse gardens, making New York into an agricultural exporter, and he has concocted just the lightweight mixture of planting

Barry Benepe, who grew up on a farm in eastern Maryland, now runs urban farmer's markets, bringing fresh produce to 50,000 New Yorkers a week. (Elaine K. Osowski)

soil and recycled Styrofoam needed to do it. (Regular dirt would collapse the roofs.)

He'd like to drape building walls with hundred-foot nylon mesh curtains of moss and ferns for air conditioning and for filtering our waste water. But one spring day Mankiewicz had settled on a more immediate project, helping school children restore an embankment on the Bronx River, when four young toughs showed up to lure the girls into the bushes.

"But they were also curious about what we were doing," Mankiewicz says. "It turned out they had captured four young garter snakes. We were looking out over one of the densest parts of New York, the Tremont area, for years one of the worst neighborhoods in the country, and here these kids had garter

snakes in their pockets just as I did growing up in rural New Jersey." Mankiewicz's boyhood home later became the Garden State Parkway, his bit of proof that the suburbs are no solution to the environmental problems of cities.

Life in the City

New York City has had three green eras. In the mid-1800s, Frederick Law Olmsted designed Central Park as the lungs of Manhattan that kept fresh air in the growing metropolis and Prospect Park as the hub of Brooklyn with grand avenues spoking outward. Many urban planners fought for green space in that era as relief from the Dickensian gloom of the Industrial Revolution. These parks allowed the common man and woman to walk through landscapes formerly reserved for mansions. But they served another unplanned purpose, apparently lowering the outbreaks of typhoid fever by 65 percent by purifying the water, according to Mankiewicz. Eco-advocates today constantly go beyond the aesthetic value of greenery to justify its place in the competitive urban setting.

In the 1930s, Robert Moses began his imperious career as the city's park commissioner—public works czar, really—vastly changing the city over the next 26 years. He tore down neighborhoods to build Lincoln Center and the United Nations, and rung Manhattan with highways. But he also laid highways out to the remote Queens beaches and built 17 new miles of beaches, part of his acquisitions that more than doubled the parkland in the city to 34,600 acres. This legacy has left New Yorkers today with 17 percent of their land devoted to playing fields, woods and marsh. Jean Gardner, author of *Urban Wilderness: Nature in New York City*, says that some of these parks retain their original ecology: the north end of Central Park and Van Cortland and Pelham Bay parks in the Bronx have changed little since the Indians. There's even a hemlock grove in the Bronx Botanical Garden that has never been cut. New York, Gardner adds, actually has the most diverse ecology of any city in the country, ranging from the rocky hemlock terrain in the Bronx to the pine barrens of Staten Island.

"New York is pretty green," adds Arthur Shepard of the New York Horticultural Society. "If it gets any greener we'd be back in the woods." He may be right. Aside from the parkland, the city has 2.6 million street trees and 700 to 800 community gardens, from pocket parks to small farms such as one in the South Bronx that produces 8,000 pounds of fruit and vegetables a year. Shepard works with the younger prisoners out on Riker's Island who cultivate seven acres of vegetables and strawberries for their own dining tables. After they get out, many find landscaping work in parks or in cemeteries, which add several more large green oases to the cityscape.

Many New Yorkers lead surprisingly green lives. Fifty thousand a week shop at 18 farmer's markets, browsing among the truck stalls of local farmers who display the richness of the regional food chain: half a dozen varieties of apples, buckets of fern fiddleheads, stacks of maroon rhubarb, tomatoes that don't feel and taste like softballs. Even more adventurous gourmets join "Wildman" Steve Brill for edible plant tours of the city's parks, noshing on weeds and berries and staring in awe at some of his mushroom discoveries. "These are oyster mushrooms," he tells one group clustered around a damp rotten tree stump in Brooklyn's Prospect Park. The mushrooms look like romaine lettuce with leaves of oyster shells. "They sell for $25 a pound at Balducci's," Brill adds, putting nature into terms Yuppies can understand.

> *M*any New Yorkers lead surprisingly green lives. Fifty thousand a week shop at 18 farmer's markets.

Seventy-five thousand people a day ride bicycles to work, including the "QB-6"—transportations's answer to the Chicago Seven—who stood trial last spring for obstructing traffic on the Queensboro Bridge in protest of the closing of the bike lane. The judge accepted their argument that they were saving lives by

---

WHAT YOU CAN DO

When in doubt about the proper way to discard a particular item, a phone call to the waste management office in your municipality is often the way to start. After that, go to the Yellow Pages. Look under "Recycling," "Scrap dealers," or "Junk dealers."

---

encouraging an alternative to the hordes of polluting cars and trucks that clog Manhattan's streets, and dismissed the case. George Haikalis, a more demure civil engineer who let his driver's license lapse over 25 years ago when he moved to the city, serves as a biking consultant of sorts with his Autofree New York Committee. What disturbs him most is that cars kill almost one pedestrian a day, making our notorious subways seem benign in comparison. His group has crafted a plan to cut Manhattan's traffic by 20 percent over four years by operating much cheaper, more frequent buses and trains; putting tolls on the remaining free bridges; and closing 20 miles of Manhattan's streets—5 percent—for walkers, trees, and sidewalk cafes.

"We're now in the third great cycle of urban greenery," says Tom Fox, director of the Neighborhood Open Space Coalition. "We have the people." Indeed, New York no longer has czars like Robert Moses or visionaries like Frederick Law Olmsted. It has broke real estate tycoons, large civil service unions and city leaders mostly good for sound bites. New York is not a city of grand designs but of vested interests, and while that might seem bad to those eco-dreamers who hope for a massive green redevelopment of the city, it is probably good for the thousands of New Yorkers creating a more environmental city from the roots up.

Looking for a Green Leader

In the 1980s, Gardner says, the city's future development underwent a watershed change with the defeat of massive construction proposals at Columbus Circle on the corner of Central Park, and in the railroad yards along the upper west side of Manhattan. In both cases architects unveiled sleek futuristic models with new towers scraping the sky; Donald Trump even envisioned a full Trump city with the world's tallest building replacing the railroad yards. But local groups, repulsed by the idea of living in more building shadows, defeated these grand plans. Trump has changed his dreams and chosen to work with six civic groups to build something more humane. Gardner even hopes he will include a salt marsh to nurture shellfish and purify the water.

The hazard of hoping for a green leader to craft an eco-city from the top can be found in Mayor David Dinkins. Reflecting the byzantine web of New York itself, the local environmental movement collects many diverse strands, from such national organizations as the Natural Resources Defense Council and the National Audubon Society to such local ones as West Harlem Environmental Action and the Lower Eastside Garden Coalition. In 1989, more than 200 groups formed Environment '89—now called Environment '91—which prepared a thorough platform for greening the city, from closing down apartment incinerators to offering low-interest loans to remove asbestos from residences, to adding a new Environmental Education Coordinator to the school system. These aren't pie-in-the sky ideas, says Gardner, who cofounded the coalition, but practical and inexpensive steps. The only financial leap would be to increase the park budget to 1 percent of the city's total, which seems reasonable for a department handling so much city land. Environment '89 then endorsed candidate Dinkins for mayor. "We quickly became the most vocal and active constituency in the city," Gardner says. "But Dinkins has come through with nothing." During this spring's budget crisis, he slashed the parks' budget by a third—much more deeply than the city's other departments—and put New York's already troubled recycling program in hibernation to save $67 million.

Recycling's Rewards

New Yorkers produce 27,000 tons of trash a day, a good chunk of which keeps piling up at the Fresh Kills landfill on Staten Island, due to become the tallest mountain on the eastern seaboard early in the next century. Sadly, says Nancy Wolf of the Environmental Action Coalition, the sanitation department has bungled its new recycling program, giving new life to a future of massive incinerators. In the early 1970s her group opened the city's first volunteer neighborhood recycling center, which gradually grew to 25 centers, and it created an apartment building collection system that spread to 400 locations. By the late 1980s, though, garbage had finally become a glamour issue, and a new law called for private haulers to recycle 50 percent of the waste from commercial buildings and the Sanitation Department to recover 25 percent of the rest by 1994. Sound great? Wolf calls the law a fatal mistake. Instead of building up

recycling slowly and surely with collection centers in all 59 community board districts, the city leapt right into curbside collection, which should have been the final step. The ballyhooed program confused the public and recovered only 6 percent of their recyclable trash. Wolf throws up her hands: "Think of the millions they've spent! If we'd had one tenth of their money we could have done anything." Today, the number of neighborhood recycling centers has dwindled to 7 or 8.

The South Bronx may not make many eco-city lists, but behind the White Castle and the Citgo station on Webster Avenue stands

WHAT YOU CAN DO

When you want to discard a large appliance, buy a new one from a company that will take the old item off your hands. Large companies like Sears generally have long-term contracts with scrap dealers to take away steel, and they will be happy to provide you the service of removing your old clunker.

the brick R2B2 plant. This morning the giant doors are open, emitting the sweet yeasty smell of beer cans, while inside you find a hill of Colt 45 cans in plastic bags, several pallets loaded with 27-pound bricks of glittering crushed aluminum and a handful of workers with ear plugs, protection from the loud waterfall of bottles constantly dropping into glass crushers. Since 1982, R2B2 has recycled 12,000 tons of material a year, buying it from people poor enough to be drawn by the prices posted at the door: 25 cents a pound for aluminum cans, 5 cents a pound for plastic bottles, 5 cents a pound for newspapers. R2B2 has paid over a million dollars so far to the locals who come with vanloads, even shopping carts full of material.

"Ten years ago, presidents came to walk through this neighborhood as one of the worst situations in the country," says David Muchnick, a lanky Harvard Ph.D. turned president of R2B2. "Fortunately, that's no longer the case." His company is part of the nonprofit Bronx 2000 development corporation founded in 1980 to revive the neighborhood. And recycling seems to be a natural answer. But Muchnick gets itchy around green talk. "This is too important to be left to sanitation officials or environmentalists," he says. "This is really an industrial development opportunity. We estimate that the collected garbage of the city has a value of upwards of $500 million a year. Scrap paper and scrap metals are already among the biggest exports from the Port of New Jersey harbor." He would like to see recycling taken out of the Department of Sanitation, which seems to be choking on the job, and given to a deputy mayor for economic development. And he'd like to see recycling companies given the same financial backing that cities grant landfill and incinerator developers.

New Yorkers produce 27,000 tons of trash a day, a good chunk of which keeps piling up at the Fresh Kills landfill on Staten Island, due to become the tallest mountain on the eastern seaboard early in the next century.

Urban Gardens

Eco-city advocates love to speak of recycling as a sustainable enterprise that can revive both city neighborhoods and the environment. For a lesson in community gardening, another eco-city favorite, stand on 8th Street between Avenues C and D in Manhattan. The north side of the block is an open field of urban hell, a homeless encampment with half a dozen scattered igloo shacks made of junk wood and sky blue tarps. A shirtless man adds more boards to his home—others stand listlessly in the sun. But the south side of the street still has occupied tenements and two gorgeous gardens. One has an elegant oval lawn shaded by a Japanese weeping pine. There's even a small bonsai garden. The other has the spirit of a Victorian backyard: a gazebo, white lawn statues, rhododendrons, a gingerbread house bird feeder. In a small playground area, children climb on a plastic honeycomb wall, gaggling like happy birds.

"Every neighborhood needs to have a community garden as a major part of it," says Barbara Earnest, head of the Green Guerillas, who have advised gardeners since 1973 when the group started its own plot at the corner of the Bowery and Houston, an intersection otherwise dominated by wino windshield washers. Community gardening dates back to the financial panic of 1893 when the mayor of Detroit opened his municipal lots to the poor; it flourished in the Depression when New York actually had 5,000 gardens and peaked in the Second World War when Victory Gardens across the country grew 40 percent of our vegetables. The current cycle began in the early 1970s when neighborhoods decided to reclaim rubble-strewn lots. For the poor, community gardens involve more than raising tomatoes, zucchini and weeds; they are social

Barbara Earnest of the Green Guerillas works in a garden on the Bowery in Manhattan, a project that sparked a movement in the early 1970s to reclaim rubble-filled land. (Elaine K. Osowski)

forums that host cookouts, school classes and Earth Day festivals. The gardens bring the disenfranchised together into block associations and nonprofit organizations, gathering the power they need to rise in the city. But Earnest thinks every neighborhood should have one, and that these gardens can begin composting organic trash and become local recycling centers. "From there we need to have Boston ivy growing on walls everywhere, and day lilies and wildflowers in every possible site." City visionaries describe vines as wonderful air conditioners and pollution scrubbers that could fill New York's many blank acres of brick and brownstone.

> *E*very neighborhood needs to have a community garden as a major part of it.

Those who don't grow their own can still buy homegrown food at the city's farmer's markets, another successful venture, which steadily grows by 8 percent a year. Barry Benepe, director of the City's Greenmarket program, got mad when he bit on a supermarket peach and almost lost some teeth. Having grown up on a farm in eastern Maryland, he knew the taste of fresh fruit, so in 1976 he opened the first market with 9 farmers on 59th Street. Today the program has 160 farmers, bakers, winemakers and fishermen, all from within 150 miles of the city, representing 8,000 to 9,000 acres of farmland. Many of these people had been running roadside stands, but now a third make their entire living from the market and another third heavily depend upon it.

Greenmarkets have limitations like anything else, floundering in poor neighborhoods. "Red Hook in Brooklyn, the South Bronx, East Harlem, they all failed," Benepe says. Greenmarkets need a lively, middle-class location, instead. Local farmers have also had no luck cracking the mainstream wholesale market out at Hunt's Point in the Bronx where the city's restaurants and groceries still buy all their produce from California, Mexico and Chile. "Hunt's Point refuses to have anything to do with them," says Benepe sadly. "They refer to their food as 'local shit.'" But he's happy to watch the program grow each year, and he'd be delighted if the city would create a permanent market site.

Clean Water

Even closer than the local farms, though, are the city's rivers, which flow almost forgotten around the boroughs except as passing scenery for the heavy traffic. Rivers once made New York great as a port and an immigrant center, but the port moved to Newark, New Jersey, and the immigrants now land at nearby Kennedy Airport. Manhattan, rung with highways and concrete bulkheads, has retreated from its shores. "The waterfront used to be a dangerous place. Think of *On the Waterfront*," says Cathy Drew, in the loft office of the River Project overlooking the Hudson from the lower west side. "How do people relate to the river? Follow me," she says, leading us to her fire escape and a seat on a wide window ledge.

Drew walks with a hitch, the legacy of an accident that ended her oceanographic career in the tropics, so she has set up scientific shop in her own backyard, a patch of the Hudson right across the road from us. It's one of the New York's uglier stretches: To the north is a prison barge with a caged basketball court on the roof and to the south is the new Amazon Club looking more like a set for "Gilligan's Island" with so many palm trees and thatch huts standing around on a cement pier. Even the river looks cheap—murky and ignored by the traffic passing below us. "This used to be a world-class oyster fishery," she says. "They had oysters as big as plates."

The Hudson was once one of the richest estuaries on earth. But Drew has found lots of life thriving below the dead surface of this scene. She has caught 32 species of fish, from eels to seahorses to bass, and she brought up blue crabs that were delicious. You can actually find lobsters off the Statue of Liberty, she adds, although you wouldn't want to eat them. She'd like to see the shoreline partly restored, a "blue-green" zone created along the concrete bulkheads with small beaches, rocky outcroppings or salt marshes that could nurture more life and remind us that we stand at the edge of a real river.

Shaping the Future

The place to end your eco-tour of New York is in the office of the Neighborhood Space Coalition with Tom Fox, who has seen it all before you and taken 14,000 slides as proof. Once he took his camera and a bottle of champagne to the Gateway National Recreation Area at Jamaica Bay to toast the demolition of an illegal construction site. "People say, you don't really mean this?" he says, showing me his sequence of shots: two concrete skeletons for apartment buildings, one dense cloud of dust, two piles of rubble that have since grown over with grass. "Yes, I do!"

But Fox isn't really an Earth First!er for New York, he's a master planner who mapped out the 40-mile Brooklyn-Queens Greenway of pedestrian and bike paths that will link the Coney Island boardwalk on the Atlantic to Fort Totten at the mouth of Long Island Sound. With 85 percent of that project now complete, he has turned his energy to the waterfront, helping the state draft a $500 million plan to create a Hudson River Waterfront Park stretching from 59th Street to Battery Park on the lower tip of Manhattan.

"Some people see cities as violations of natural systems, and they are," Fox says. But healthier cities are crucial to the health of the larger environment. More livable cities could reverse suburban sprawl. More sustainable cities could encourage regional farming and the revival of neighborhoods through gardening and recycling. And more environmentally aware city dwellers can help shape a greener future for everyone. "By the year 2000, 80 percent of the population in the United States will be living in cities," Fox says, referring to metropolitan regions that include suburbs. "The large environmental groups are concerned with *out there*. But I think we need to bring it *in here*, too."

Reprinted with permission from *E—the Environmental Magazine*, September/October 1991. Subscriptions $20/year; P.O. Box 6667, Syracuse, NY 13217; (800) 825-0061.

The Eco-City Movement Offers a Greenprint for Downtown U.S.A.

By Francesca Lyman

Enter the heart of this big city, and be dazzled. Solar-powered buildings glisten in the sun. Trees sprout from rooftops. Verdant community food gardens grow over what used to be rubble. There is no traffic din or smog, because what cars remain whirr past on electric power, beside a lane of softly humming bicycles. Urban creeks flow into waterfalls and spectacular foundations, causing cool breezes to waft toward a nearby sidewalk cafe, where a sign reads, "Lunch is not a meal, it's a sensuous experience."

If these visionary images seem like something out of the dream journals of futuristic planners, you're right—but only partly. From the bicycle-friendly streets of Berkeley, California, to the community gardens of the South Bronx, a new breed of activists has taken up a cause largely left out of mainstream environmentalism: urban ecology. Composed of a motley mix of architects, planners, landscape designers, community organizers and others, this new "eco-city" movement is laying the groundwork for cities to become healthier and more livable. More than 700 people from around the world attended "The First International Eco-City Conference" in Berkeley in February 1990 to hear, among others, Richard

Register, president of Urban Ecology, say that "no ecologically healthy city exists today. We do, however, see hints of the eco-city in today's solar, wind and recycling technologies; in urban gardening and tree planting projects; and in individuals using foot, bike and public modes of transportation."

Man's Monoliths

Cities have long been thought of as the antithesis of nature. The English poet William Cowper once wrote, "God made the country, and man made the town." As human work has proliferated, often without thought to God's country, cities have come to seem like heartless, carnivorous creatures devouring the rest of the planet. Take Los Angeles. Please. To Peter Berg, author of *A Green City Program for San Francisco Bay Area Cities and Towns*, the City of Angels is a kind of parasitic organism that "drains water from Northern California, extracts coal for electric power from the Four Corners area and ships in liquefied natural gas from Indonesia." With cities competing for increasingly scarce supplies of water, energy and food, he wonders if urbanites will "be vulnerable to chaotic shortages and supply break-downs."

*F*rom the bicycle-friendly streets of Berkeley, California, to the community gardens of the South Bronx, a new breed of activists has taken up a cause largely left out of mainstream environmentalism: urban ecology.

"There is a monolithic development in cities around the world. They are asphalted, human-dominated, exploitative, polluted, unhealthy habitats," writes Dr. Rashmi Mayur, director of the Urban Environment Institute in Bombay, India, in the book *Green Cities.* "Only catastrophe awaits such a system of disharmony."

Yet environmentalists cannot simply abandon cities to such grim fates. Too much of humanity has pinned its hopes on urban life. In 1950, less than 30 percent of the world's population lived in cities and towns of 25,000 or more, according to the World Watch Institute. By 2000 that figure will swell to 50 percent, and it will top 75 percent in Latin and North America, Europe and East Asia. If these metropolises drain away Earth's resources, everyone will be the worse for it.

The American Dream of suburbia was thought to be the antidote to cities. And we have pursued the dream in droves. "Suburban residents—considered part of the urban population—have swelled from 35 million in 1950 to more than 120 million in 1990. Today's suburbia extends for dozens of miles from the old city center, often covering an area 10 to 20 times larger than the traditional pre-World War II city," says a recent issue of *The ZPG Reporter* from Zero Population Growth. This dispersed, low-density pattern of development turns out to waste far more land and generate far more air and water pollution than the traditional pattern of high-density urbanism. Eclipsed by so many other pressing environmental problems during the 1970s and 1980s, the issue of sprawl is again rearing its ugly head. The World Wildlife Fund/Conservation Foundation concludes that "the automobile-dependent sprawl mandated by many suburban and rural communities is destroying America's cities and countryside to the detriment of environmental, cultural, social and economic values."

Urban planner Will Abberger of the Conservation Foundation says that rapid suburbanization has transformed the Chesapeake Bay, for instance, "from one of the most important fisheries in the country to the one of the least." The Chesapeake Executive Council's *2020 Report* notes that paving over larger areas of land creates more stormwater runoff and concentrates its toxicity, since the water no longer filters through the soil. And with pavement came real estate. Much of the farmland lost since the 1970s is due to suburban development, says Bob Wagner of the Massachusetts American Farmland Trust. Farmers realized, says Wagner, that "You can make more money growing houses than cows." The American Farmland Trust notes that every year, the United States loses more than one million acres of productive farmland—an area larger than Delaware—to urban sprawl.

A New Metropolis

Throughout history cities have been conceived as built environments separate from the natural landscape, but today's masses of con-

WHAT YOU CAN DO

If you're cleaning house and you have discardable items, consider contacting the Salvation Army. The number is listed in the White Pages—and there are more than 7,000 branches around the country.

crete, stone, metals, plastic and increasingly synthetic materials need not be seen as our only fate. In a recent brochure, the Trust for Public Land, a conservation group dedicated to adding green space to cities, wrote, "It's time to stop thinking of our cities as one place and nature as someplace else. Our urban centers and edges host the vibrant variety of our culture. We should not have to think of them as places to escape. Bring on the green spaces, we say: The pocket gardens, the community parks, the river corridors and the tree-lined boulevards. For those serious about saving the environment, the cities are a logical place to start."

What's new about the eco-city movement is that planners, architects, designers and others are beginning to look at urban development in a unified, whole-systems way. For example, people are rethinking the whole notion of transportation. Richard Register, known for his statement, "Transportation is what you have to do to get where you want to be," says we ought to be building for "access by proximity, not transportation." So architects and urban planners are returning to old-fashioned ideas of "mixed-use" town planning in which homes, shops and offices are all built closely together to avoid the need for transport. At "The First International Conference on Auto-Free Cities," held in New York City last May [1990], alter-

native transportation advocates, dressed in T-shirts that read "one less car," pushed everything from slowing traffic to developing human-powered vehicles.

Eco-city activists have learned to talk economics in defending the place of greenery in the urban matrix. A city street without trees loses the benefits of shade, beauty and purified air. In sheer fiscal terms, "the goods and services," provided by an ecosystem are lost. Robert Costanza and Lisa Wainger in *Mending the Earth* write, "We know what it costs to build and run sewage treatment plants. Since natural ecosystems perform these same services for free, they are worth at least the amount we would pay for corresponding human-produced services." And Tom Fox of the Open Space Coalition in New York adds that parks and open spaces are economically critical to cities "because they have always added to real estate values."

W here you have strong parks, you have strong communities.

The very notion of thinking of the environment as a luxury is impoverished, says Morgan Grove, director of Yale University's Urban Resources Initiative, a project set up three years ago to work primarily with inner-city youth in Baltimore, New Haven and Detroit, creating gardens and restoring parks. "In places without trees, you feel an absence affecting people's lives," he says. Rather than addressing poverty and the environment separately, his group seeks "to link urban revitalization with environmental restoration in order to address the needs of urban residents and break cycles of social and environmental deterioration."

Grove worked with one group of children in Baltimore who were quick to recognize the many beneficial qualities of trees. "So one child's response," he says, "was 'How come there are no trees where I live?' " And his group discovered: "Where you have strong parks, you have strong communities."

Such thinking needs to filter more deeply into the urban mind. "Most of the people who design and build cities have successfully isolated themselves from the natural sciences and make no attempt to cultivate even a superficial relationship," says Gary Moll, a vice president of the American Forestry Association. Perhaps, too, city dwellers have come to expect the benefits of what was once "a rare and privileged way of life," suggests Peter Berg of Planet Drum. But this rarified life, denuded of the natural world, has itself come to seem incomplete. An urban acre of green space can attract a surprisingly large and diverse variety of wildlife drawn to the habitat. It can make city dwellers more aware of their sense of place, as well as the essential ecology and natural history of their home turf.

The seedlings of Eco-city U.S.A. are already growing. In Berkeley, activists are restoring an urban creek long buried in an underground culvert; they have even introduced crayfish and trout to waters once locked in an endless night. In Baltimore, kids can get a free ticket to the city zoo for the price of recycling a few bottles. In Boulder, people can ride the bus free if they promise they won't drive their car that day. In Los Angeles, there's a proposal to require that electric cars be on the road by the year 2006. This city of sprawl, so often held up as an example of what's unsustainable about urban life, even held its own "ecological cities" conference in May [1991]. These seedlings will take time to flower, but what a glorious sight we will have when the old barrier between nature and towns has fallen.

Reprinted with permission from *E—the Environmental Magazine*, September/October 1991. Subscriptions $20/year; P.O. Box 6667, Syracuse, NY 13217; (800) 825-0061.

Winding through Pennsylvania's Pocono Mountains, the Lehigh River gushes and froths. (Courtesy of Whitewater Challenges, Inc., Whitehaven, Pa.)

THE CLEAN WATER ACT TURNS 20

By Jeff Rennicke

In 1969 Cleveland's Cuyahoga River did something water isn't supposed to do: It caught on fire. Hot slag accidentally spilled from a plant, and the surface of the river, which was more oils, chemical foam and debris than water, was set ablaze. Although the spectacle of a burning river made international headlines and turned the city of Cleveland into the butt of merciless jokes, the bitter reality was that despite federal water-quality actions, many rivers, lakes and streams were dying. In early 1972, the Senate Public Works Committee, in a moment of uncharacteristic candor, admitted that "the national effort to abate and control water pollution has been inadequate in every vital aspect."

Something had to be done, so on October 18, 1972, the Federal Water Pollution Control Act, better known as the Clean Water Act, was born. In the 20 years since, no more rivers have gone up in flames. But are the waters cleaner than in 1972? Has the act had an effect in the backcountry or has it been strictly an urban concern? What sections of the law need to be strengthened?

Conservation groups are asking these and other tough questions as they mark the anniversary of the act and prepare for what is expected to be a hard-fought reauthorization battle in Congress.

"The Clean Water Act has yet to achieve its goals and objectives," says David Dickson of the Izaak Walton League. Dickson can make such a bold statement because those "goals and objectives" were specifically laid out in the act itself. In addition to its objective to "restore and maintain the chemical, physical and biological integrity of the nation's waters," the act set out to eliminate the discharge of any pollutant into U.S. waters by 1985. All the nation's waters were to be safe for fishing and swimming by July 1, 1983.

Successes and Failures

If judged strictly by the deadlines, the act has failed. The Environmental Protection Agency (EPA) says more than 17,000 U.S. bodies of water remain heavily polluted, with many of them unsafe for fishing or

185

swimming. In the Great Lakes alone 68 percent of the coastal areas did not meet the "fishable" criteria in a recent survey. Throughout the country, more than 700 fishing restrictions remain in place.

There are success stories, however. Initially the act focused on "point source pollution"—the obvious offenders being municipal sewage treatment and industrial discharges—and some waterways have shown spectacular results. Thirty-two sewage treatment plants on the Chesapeake Bay have recorded a 98 percent reduction in chlorine releases since 1984. Along Wisconsin's Fox River, which has more mills per mile than any waterway in the world, industrial discharges in 1980 were one-tenth of their preact levels. The Fox, once on the EPA's infamous "Dirty Dozen" list, now has nesting bald eagles, trout and even a few canoeists.

I ndustrial releases of toxic substances still total 362 million pounds a year.

Huge problems remain, however. Industrial releases of toxic substances still total 362 million pounds a year, and the EPA estimates it will cost industry another $83.5 billion to comply with sewage treatment standards through the year 2008. Still, most experts agree that the control of point source pollution is the success story of the Clean Water Act's first 20 years.

As the cleanup of obvious pollution sources proceeded, less obvious and more problematic issues began to turn up, often in surprising places. For instance, Siskiwit Lake is surrounded by unbroken forests

> ## WHAT YOU CAN DO
>
> For sink and drain stoppages caused by hair, food or other solid matter, try boiling water, snaking the clog with a metal line or using a plunger. All are better than toxic chemical alternatives.

and sits like a blue jewel near the heart of Isle Royale National Park in Lake Superior. In the mid-1970s, an EPA team in search of pure, pollutant-free water to act as a baseline in water-quality tests found snow there with nearly five times the level of contaminants as urban snow, and fish with dangerously high PCB levels. Siskiwit was one of the first documented backcountry victims of NPS, or "nonpoint source pollution."

NPS is pesticides washing off a farmer's field into a creek that feeds into the Boundary Waters Canoe Area. It's Colorado's Arkansas River, a popular whitewater run, tinted orange from a leaking abandoned mine. It's landfills that leach into the New Jersey Pine Barrens. It's brown streams choked with silt because poor logging and grazing practices triggered erosion. It's the single biggest threat to our nation's water quality, whether in an urban reservoir or a backcountry trout stream.

Because there's no single, easily identifiable source, combating NPS is like boxing with shadows. A 1987 amendment to the act (Section 319) was a first step, but its voluntary compliance programs haven't worked. With reauthorization of the act coming up, conservation groups are expected to push for the same kind of programs that worked on point source polluters.

Wetlands Neglect

When it comes to our mistreatment of water and water-based ecosystems, however, the neglect of wetlands stands out. A survey by the U.S. Fish and Wildlife Service estimates that half our original wetlands, more than 117 million acres, were lost between 1780 and 1980. They were dredged, drained, filled and paved over at the rate of 60 acres an hour. Florida alone has lost 9.3 million acres. In California and Ohio, which have both lost more than 90 percent of their acreage, the word "wetland" has ceased to exist.

W hen it comes to our mistreatment of water and water-based ecosystems, the neglect of wetlands stands out.

Once we passed over them as swamps, lowlands or bottoms that were of little environmental significance. Today we know wetlands provide important flood control, migration and breeding grounds for birds, water quality benefits, recreation and wildlife habitat. Over one-third of the plants and animals on the endangered species list depend on wetlands. Yet for all their proven ecological importance, only one federal program provides a regulatory mechanism for wetland management: Section 404 of the Clean Water Act. And even that doesn't do much.

Section 404 sets up a procedure for permits that must be obtained before a wetland can be filled. Period. It does not provide a system of protective designation, nor does it apply to the draining, channelization or dredging of wetlands. Most agricultural uses are exempted even though farming is responsible for up to 85 percent of the wetlands loss. The few wetlands victories that have occurred are the result of the EPA's veto power over permits issued by the U.S. Army Corps of Engineers. Two particularly notable examples are the defeat of the Two Forks Dam on the South Platte River near Denver, which would have disrupted wetland habitat for the endangered whooping cranes downstream, and the halting of a planned reservoir on Virginia's Ware Creek that would have threatened great blue heron rookeries.

Still, we're losing wetlands at a rate of 290,000 acres a year. To stem this loss, conservationists will be trying to strengthen Section 404 to regulate all wetland-altering activity, to allow more public input, and to include wetland protection as a specific goal of the Clean Water Act. In light of the Bush administration's recent attempt to "redefine" wetlands, which would have opened 50 percent of the remaining wetlands to development, many groups also want a rededication of the act's scientific definition of wetland ecosystems.

Making Things Better

The constant need for vigilance over details like the administration's interpretation of words points to what may be the act's major **shortcoming**: its emphasis on trou-

We're losing wetlands at a rate of 290,000 acres a year.

bleshooting. "In practice the act has always been reactive rather than proactive," says the Izaak Walton League's Dickson. "But a close reading makes it clear that the act is supposed to also preserve water quality in areas where it's still pristine."

And where is most of the remaining pristine water found? In the backcountry, but keeping it clean is not as easy as it sounds. Funding the cleanup of an already polluted waterway gets more play in the media and on the political front than the allocation of money to maintain an apparently healthy wilderness waterway. Yet it's easier and less expensive to keep an already clean waterway clean. Conservationists stress that we shouldn't wait until there's trouble. Although "antidegradation" and "antibacksliding" amendments were added to the act in 1987, environmentalists will push for a stronger program of maintenance and protection this year [1992].

And there is more. Groundwater protection, a classic out-of-sight, out-of-mind subject, is vital to the backcountry since it provides 40 percent of the flow in our rivers and streams and is an important source for lakes and wetlands. Interpretation of the act by state and federal agencies has focused, conservation officials say, too closely on the chemical quality of water and should adopt an ecosystems approach to water quality. There is a need to sort out little-used clauses such as Section 401, which regulates the effects of federally licensed projects on streams. With more than 200 federally approved hydroelectric dams coming up for license renewal in the next few years, clarification of this section is vital to backcountry water quality.

The original optimism of the Clean Water Act and its ambitious deadlines has evaporated, leaving behind the reality of a difficult, expensive task, the scope of which was unfathomed 20 years ago. Hope remains, though. In the words of former Senator Edmund Muskie, a supporter of the original bill, "In celebrating the Clean Water Act, we do so not as a symbol of the success we have achieved, but as a symbol of the work that still needs to be done."

From *Backpacker,* June 1992. Reprinted by permission.

River Revivals

By Annette Wysocki

For much of American history, our rivers have been considered workhorses for industry rather than environmental assets to be enjoyed and preserved. "In the interests of national welfare, there must be national control of all running waters of the United States," the dam-minded policymakers of President Franklin Roosevelt's administration advised in 1934. President Harry S. Truman's Water Policy Commission furthered that attitude in 1950: "The American people are awakening to the new concept that river basins are economic units." Pristine, free-flowing waters were being dammed, channeled and poisoned at a record pace, all in the name of progress.

> *Although it's just about impossible to truly restore a river, you can make it canoeable, hikeable and enjoyable.*

Then came the 1970s and a change in attitude. People were tired of putrid rivers that changed colors faster than a mood ring. Federal legislation, specifically the Clean Water and Wild and Scenic Rivers acts, helped restore desecrated rivers and prevented clean ones from becoming foul.

Although it's just about impossible to truly restore a river—that is, return it to its pristine, pre-Columbian condition—you can make it canoeable, hikeable and enjoyable. Following are examples of some of the most successful cleanup efforts. Go and enjoy them.

The Nashua

Before the Pilgrims arrived, the Nashaway Indians fished in a pristine waterway they called "the river with the beautiful pebbled bottom." By the 1960s, the Nashua River, which flows through north-central Massachusetts and southern New Hampshire, was one of the most polluted waterways in the country, little more than a 56-mile sewer filled with waste from the paper mills, plastics factories and other industries lining its banks. The river varied in color, depending on what had been dumped that day, and the stench was so strong that some residents of Hollis, New Hampshire, complained the odor kept them awake at night.

One of the first and strongest proponents for saving the Nashua was Marian Stoddart, a seasoned community activist living near Fitchburg, Massachusetts, whose 30 years of activism earned her the nicknames "Mother Nashua" and the "Clean River Lady." Stoddart formed the Nashua Cleanup Committee in 1965, despite a declaration from the U.S. Army Corps of Engineers that the river was dead and no fish or wildlife could survive in its waters.

The committee quickly grew in numbers and was a key force in the passage of the state's clean water act. As part of its lobbying effort, the group gave then Massachusetts Governor John Volpe a jar of nasty Nashua River water to keep on his desk until the bill became law. That was the easy part. The real battle came in the late 1960s when the committee had to convince industries and municipalities along the river that they would benefit from the expensive changes needed to meet higher water quality standards. Some paper companies complained that the cost of waste treatment would cause layoffs. With the help of federal and state funds, however, wastewater treatment plants were built and industries were able to comply.

By 1969, the committee had grown into the Nashua River Watershed Association and had enlisted civil engineers, landscape architects and attorneys to draw up a comprehensive plan for the entire Nashua River watershed. The plan went far beyond just getting the muck out of the water; it called for construction of wastewater treatment plants, greenways and protection and restoration of wildlife and plant habitat. In the past two decades, the group has helped preserve more than 70 miles of riverfront and some 6,000 acres of Nashua watershed. Major land acquisitions include the Squannacook Wildlife Management Area, which has the best trout streams in Massachusetts; the Oxbow National Wildlife Refuge, an important wetland sanctuary for migratory waterfowl; and Bolton Flats Wildlife Management Area, home to 300 species of animals. Although the Nashua still has some water quality problems, the Massachusetts Department of Environmental Management now considers it a "scenic river."

The Potomac

The Potomac begins in West Virginia and flows 383 miles through Virginia, Washington, D.C., Maryland and into the Chesapeake Bay. Along the way it winds through scenic Appalachian mountains and rolling Piedmont hills and encompasses a coastal plain estuary that's 11 miles wide where it meets the Chesapeake. But for decades, sulfuric acid from coal mining operations left some northern sections devoid of plant and animal life. Farming above Washington caused so much erosion that experts predicted some sections of the river would be filled in within 50 years. The relentless suburban sprawl of the nation's capital took a heavy toll as well. By the 1960s, the Washington area had grown to more than two million people and some 500,000 acres of farmland had become housing developments. Inadequate waste treatment plants raised coliform bacteria levels to extraordinarily high levels, making fish kills common and rendering the Potomac River off-limits for most recreational activities.

Although an Interstate Commission on the Potomac River Basin was established in 1940, it had limited authority until 1970, when Congress put the group in charge of overseeing all water-related matters in the Potomac basin. At that time, the commission implemented a comprehensive water quality monitoring program, drummed up federal and state funding for wastewater treatment plants and helped coordinate community involvement in river cleanup activities. These efforts, along with the help of thousands of volunteers, started yielding results in the 1970s. During that decade, $1.6 billion was spent on wastewater treatment plants. Fish, including largemouth bass, slowly reappeared. Today, the

Potomac is one of the most popular recreational rivers in the Mid-Atlantic.

The Pigeon

For decades, Appalachian Trail hikers passing through the Great Smoky Mountains tried to stomach the stench while crossing the bridge over the Pigeon River at Waterville, North Carolina. Canoeists and kayakers stared longingly at the surging rapids, wishing there were some way to run them without becoming nauseous. Now that Champion International Corporation, the company whose pollution has been causing the big stink, is cleaning up its act, the long-suffering Pigeon may soon return to its naturally healthy state.

Not long after Champion opened its Canton paper mill in 1908, the Pigeon became worthless as a recreational source. The river enters the mill clear and odorless but emerges the color of coffee, smelling like an outhouse. Organic waste matter from the papermaking process was robbing the river of oxygen. Water temperatures rose to the point where few of the once-plentiful trout were able to survive; those that did contained dangerous levels of dioxins.

In 1981, Champion's permit to discharge wastewater into the Pigeon expired, and concerned citizens lobbied for change. Fearing a loss of jobs if the mill closed, the North Carolina Environmental Management Commission drafted a new permit with weak and unenforceable limits on water temperature and color. Officials in Tennessee, the river's destination below Canton, promptly sued North Carolina for failure to comply with the federal Clean Water Act. Then in 1985 the Environmental Protection Agency got into the act, and under federal pressure Champion unveiled plans for a $250 million antipollution modernization plan to be completed in 1993.

Improvements in quality are obvious, but everyone is taking a wait-and-see attitude before giving the cleanup effort their unqualified blessing. If all continues to go well, a nightmare could become a river runner's dream.

The Willamette

The Willamette River's headwaters are located in the Cascade Mountains in the Willamette National Forest. The river, which has been called the "lifeblood of northwestern Oregon," winds through forests and farmland until its confluence with the Columbia River in Portland. Anyone who fishes the Willamette's clear waters or hikes its protected greenways these days may be surprised to learn that it almost choked to death on pollution during the first half of this century.

Waste from paper mills and food processing, in addition to raw sewage, made the Willamette (especially the stretch from Eugene to Portland) one of the nastiest rivers in the Northwest. With rafts of sludge floating on the water and slime coating its banks, the river was virtually useless for any kind of recreation. Perhaps most tragic was the resulting decimation of the Willamette's famed chinook salmon population. By the 1960s, only a few of the fish could survive the 160-mile swim from the ocean to spawning beds upstream without suffocating in the oxygen-depleted water.

Although clean-up groups had been trying since the 1920s to gain widespread support, it took a 1961 television documentary by newscaster Tom McCall to raise the ire of Oregonians statewide. Restoring the river became a major issue, as well as a campaign platform for McCall, who was elected governor in 1966.

What followed was a series of strict water quality and river conservation laws that made Oregon one of the most environmentally progressive states in the nation. Over the past 25 years, industry has complied with the state-monitored waste discharge standards, municipal wastewater treatment plants have been built for riverfront communities, fish ladders have been constructed to help the salmon on their journey, fish breeding programs have been implemented to replenish depleted stock and 255 miles of greenway have been acquired, enough to create a large riverfront park system.

The Environmental Protection Agency called the Willamette's restoration a "model for the nation." Water quality has increased dramatically, the salmon are back, and the river and its greenways are a recreational oasis.

From *Backpacker*, June 1992. Reprinted by permission.

 EARTH CARE ACTION

Students Take the River Test

By Deborah Simmons

What do students along the Ganges River in India, the Murray-Darling in Australia and the Rouge in Michigan have in common? They and thousands of others around the world belong to the Global Rivers Environmental Education Network (GREEN). Developed by Dr. Bill Stapp of the University of Michigan's School of Natural Resources, GREEN teaches students to nurse our ailing rivers. "One out of four hospital beds in the Third World is occupied by people suffering from water-borne diseases," he says. "Eight-five percent of the world's population lives around rivers."

The GREEN Project teaches students how to monitor the water quality of the rivers in their own communities. Saginaw, Michigan, schools, for instance, hold a "River Walk Day," with canoe races and other water activities. But with their monitoring program, students found that the river had unsafe coliform levels (a type of bacteria found in human and animal feces), a sign of sewage contamination. "The students took their data to the city council and suggested an alternative site for the river walk," Stapp says. "But it didn't end there. They discovered why the fecal coliform levels were so high—the city's storm water and sewage ran through a combined sewer overflow system." Whenever rainstorms overloaded the system, untreated sewage would flow right into the river. Due to the student efforts, the city council placed a bond issue for a revamped sewer system on the ballot. "The students helped pass the bond issue. They wrote an eight-page section in the newspaper to educate their parents, and they went campaigning door-to-door," he adds. Saginaw is now building a separate sanitation system.

---

*E*ighty-five percent of the world's population lives around rivers.

GREEN programs now exist in 35 countries, so students can learn not only from their own rivers but from others in their network. Eastern and Western Europe are beginning to look at the whole Danube River system as a common resource. Egypt and Israel are planning a joint water monitoring program. "It's a people-to-people exchange of thoughts and ideas," Stapp says. "It's built on the idea that rivers know no political boundaries." (Contact The GREEN Project, School of Natural Resources, Dana Building, University of Michigan, Ann Arbor, MI 48109; 313-764-1410.)

Reprinted with permission from *E—the Environmental Magazine*, January/February 1992. Subscriptions $20/year; P.O. Box 6667, Syracuse, NY 13217; (800) 825-0061.

EARTH CARE ACTION

Arm-of-the-Sea Theater

By Rima Nickell

You don't forget an Arm-of-the-Sea Theater show. From the opening drumbeats to the curtain call, when the puppeteers emerge from backstage and the players doff their masks, the presentation showers your senses with vibrant colors, in addition to provocative textures, images and sounds.

Since 1982, Arm-of-the-Sea Theater has dramatized Hudson River ecological and social issues at numerous locations and events from Albany to New York City. For cofounders and directors Patrick Wadden and Marlena Marallo, of Maiden-on-Hudson, the increasing visibility and popularity of the troupe's performances are the fruition of nearly a decade of visionary hard work.

Arm-of-the-Sea shows combine ten-foot-tall puppets, masked actors, animated graphics and unusual music and sound effects with dialogue and poetry; the productions astound,

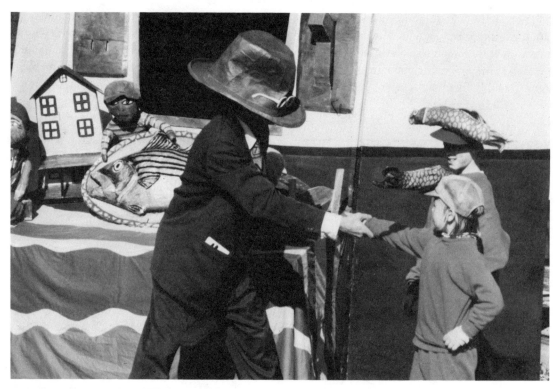

Colorfully masked actors, animated graphics and unusual sound effects are the ingredients for an Arm-of-the-Sea production. (Rima Nickell, courtesy of Arm-of-the-Sea Productions, Inc.)

amuse and challenge your imagination, intellect and conscience. Their subjects range from basic Hudson River ecology and solid waste recycling to tropical deforestation.

Like the lower Hudson, where the mixing of river and ocean forms the fecund estuary that is, literally, an arm of the sea, the namesake theater company also blends diverse elements—visual and theater arts, passionate concern for the environment and social activism. It's entertainment with a cause, environmental education with pizzazz.

A Mystical Love

Wadden and Marallo began as idealistic young artists searching for a way to make a difference through their art. Neither had formal backgrounds in environmental science or theater arts. She was an artist-designer, he a poet, sculptor and sailor. But they shared an appreciation of beauty, concerns for social justice and ecological balance, and a deep, almost mystical, love of the Hudson River.

Marallo, the artistically gifted child who "drew" her way through elementary school, sketching and doodling incessantly, awakened to the river's mystique during family drives in the Hudson Highlands. In high school in Somers, New York, she participated in a "Walk for Water," organized by the Hudson River Sloop Clearwater. "Kids from all different schools," she recalls, "would raise money to clean up the river by getting sponsors to donate a penny a mile. It seemed like peanuts, but we'd end up raising several hundred dollars."

Kids from all different schools would raise money to clean up the river. It seemed like peanuts, but we'd end up raising several hundred dollars.

While studying at State University of New York at Purchase, Marallo became inspired by the work of woodcut printmaker Antonio Frasconi. "He is a great communicator," she says. "He is an artist who bridges the personal and the political. His prints are like visual poems that spark our humanity."

She and some fellow students rented a house on the river in Cold Spring across from Storm King Mountain, a spot she calls "one of the most beautiful places on earth." There she designed stained glass windows and ceramic tile mosaics and created, in Wadden's words, "exquisite little woodblock prints," all inspired by the Hudson River.

When a local chapter of the Clearwater organization docked its sloop *Woody Guthrie* in Cold Spring to facilitate a litter cleanup of Little Sandy Point Beach, the girl who had "walked for water" in high school became an activist and organizer. "Here was this beautiful beach," she recalls, "with more garbage on its shore than any one person could possibly pick up. I had an idea to throw a big party. I made woodblock prints of the river landscape for an invitation and personally handed them out to people on several crowded weekends before the sloop's arrival. I was surprised how many people turned out and how much garbage we picked up together. It was a great time: All

kinds of people took an interest in preserving the safety and beauty of the beach. It was work that felt meaningful and necessary."

A self-taught poet and sculptor, Wadden developed "river love" early in life growing up in the upper Mississippi River town of Winona in southeastern Minnesota. Despite the lack of music or art programs in Winona's small school system, Wadden says that "two or three maverick teachers gave me a good grounding in humanities and literature." His ecological awareness grew through Boy Scout activities, frequent camping trips to nearby wooded areas and association with American studies teacher Jim Mullen, who was "way ahead of his time in ecological literacy."

Coming to the Hudson

After high school graduation in 1971, Wadden continued his education in public libraries and national parks. He explored the Rocky Mountains, lived and worked in Alaska and North Dakota and traveled extensively "trying to make sense in a country that seemed on the edge of revolution over the Vietnam War."

During the same period, Wadden spent extended periods in the Atlanta home of his elementary school printmaking teacher, Barbara Brozik, and sculptor Max Green, his first exposure to people "living in the arts," an experience that was to profoundly influence his life. "They didn't just make a living, but lived out their expression in the arts. They weren't dilettantes. It was their passion."

He returned to Winona and lived for five years on a Mississippi houseboat, taking college classes in literature and philosophy, writing poetry and carving wood. He worked as a commercial fisherman, apple picker and tug-boat crewman.

A chance introduction to the sloop *Clearwater* during a late 1970s visit to Albany was to bring him permanently to the Hudson Valley. Charmed by the *Clearwater* and her mission, he signed on as a crew member, then as first mate, from 1979 to 1981. After another year crewing for the whale research vessel *Regina Maris*, a square rigger out of Boston, he began soul-searching. "I was looking for ways to coalesce my 15 years of experience. I had bad eyes so I couldn't get a captain's license. I'd lived and worked on the water, but if I'd continued it would probably have been working on other peoples' yachts, and I wasn't interested in that."

So, late that summer, Wadden returned from Vermont to the valley, paddling down the Hudson by canoe with a dream that linked his life with Marlena and has guided their lives since.

"I was looking for a creative way to reach people the environmental movement was failing to reach. It didn't seem to take in the people who were most affected by social-ecological decisions being made in the region in terms of land use and the fallout from industrial pollution. So the whole basis of what I wanted to do was not within the paradigm or normal confines of the environmental movement."

"I'd seen Bread and Puppet (a large-size-puppet theater company based in Vermont) and thought that kind of approach would lend itself to the environmental issues and audiences I wanted to address then, particularly people in the river towns." Wadden thought that a big spectacle connected with the annual Pumpkin Sail of the *Clearwater* could attract those audiences, and that the huge characters would stand out even in such a large-scale event.

So, with logistical support from Clearwa-ter, he begin work on a riverfront pageant for the 1982 Pumpkin Sail. He had been impressed by Marallo's woodcuts and enlisted her to design the visual elements. In the *Woody Guthrie* boat shed in New Hamburg, using scrap materials, they created the first giant characters that have become the hallmark of their productions.

Morality Play

Entitled Pumpkin Seeds, the pageant was, in Wadden's words, "an outrageous morality play about the Genie of the Atom, the Pumpkin Earth, the villain Mr. Avarice and Rip Van Winkle, whom the kids woke up to stop a greedy monster from destroying the whole world."

Nine years and many performances later, they still recall the first experimental performance with astonishment and excitement. They reminisce that it was a fantasy come to life. "Who would imagine in this day and age that there's still a boat like the *Clearwater* sailing the Hudson, stopping at towns all along the way from Albany to New York City, with banners flying and the ship's crew playing instruments and singing songs?

"We did parades through the streets before the show started and got all kinds of kids to follow. You can imagine the spectacle it created on South Street in Manhattan. A hundred kids carrying the yards and yards of patchwork-blue cloth representing the Hudson River. Hundreds of people came."

The show was a catalyst. They had found their niche. They had established the basic format, style and purpose that would uniquely characterize later productions. They were creating, as their playbills proclaim, "a touring folk theater that performs on behalf of the entire biotic community." And they had found each other (and they later married).

WHAT YOU CAN DO

Turning off the water while you shave and turning it on again to rinse the razor is a thoughtful way of conserving water.

That winter, with a small foundation grant, they began work on *Scenes from the Invisible World* and *Estuary Tales.* They believed they had found a medium for transforming complex ecological concepts into messages intelligible to audiences of varied ages and backgrounds, and making them accessible to those most affected by environmental decisions. They continue to build shows around the morality play format and the use of allegory. "We also wanted to speak to a wider cross-section of the river town populations and make ecology comprehensible to them. So we did the narratives in both Spanish and English."

Their association with Hispanic Americans inspired a visit to Central America, where they communicated and made friends through a Spanish-captioned photo album of *Scenes from the Invisible World.* When they returned in 1984 to Wadden's home town, they created *Give Us This Day,* based on their Central American experiences and supported by a Minnesota State Arts Council grant.

By then, a baby was on the way, so they returned to the Hudson Valley where their son, Noah, was born in the spring of 1985. That year they worked as a theater-in-residence at Oakwood School in Poughkeepsie and re-staged *Give Us This Day* with high school students. This work brought into greater focus their interest in the concept of the arts in social change. They remember struggling to work on

their local and regional levels within the context of the global situation.

Ecopuppets

Marallo and Wadden's puppet technology reflects their ecological sensitivity. They use mostly inexpensive and recycled materials, transforming, for example, large-appliance cartons into colorful giant fish and clothes-dryer exhaust hoses into glistening jumbo earthworms that are manipulated with dowels. They sculpt the giant puppet heads in clay, then overlay these forms with strips of brown grocery-bag paper, dipped in wallpaper paste. After the papier-mâché shell dries, they paint the heads and then rig them to backpack frames or poles.

In this age of high-tech electronic media, their approach may seem primitive, but Marallo says, "We're drawn to this kind of visual theater because it is very direct: There's a powerful immediacy about it, and we can do it on a relatively low budget."

Since 1982, Marallo and Wadden have created a dozen original productions dealing with a range of Hudson Valley environmental and social issues. Recent works include *A Silver Swarming,* about PCB pollution and the plight of the striped bass; *The Secret Lives of the Estuary,* on the ecological importance of plankton (the river's unseen inhabitants); and *Force of Habit,* which deals with solid waste problems and promotes recycling.

For the 1989 Pumpkin Festival, Clearwater commissioned the company to create *The Great Tablecloth,* a performance poem featuring puppets, paintings and drums that dramatizes the plight of displaced and homeless people in the Hudson Valley. During the 1991 festival, they presented *Feast of Life,* an updated view of the Hudson River PCB situation.

Still retaining the characteristic of community or folk theater in its accessibility to a wide diversity of people, Arm-of-the-Sea Theater now employs composer-percussionist Brian Farmer to compose and play the music and a core of six professional actor-puppeteers who tour and perform year-round. Last season they presented over 80 performances throughout the region at schools, parks, festivals and cultural events, ranging from 4-H fairs and local Heritage Day celebrations to the Lincoln Center's Out of Doors Festivals.

Besides creating and performing their shows, Arm-of-the-Sea members conducted workshops for students, teachers and others, tailoring the subject and activities to the particular needs of each group. Marallo and Wadden also continue producing woodcut prints for an annual print sale at their studio in Malden.

The company's newest work, *The Water Tree*, dramatizes the interrelationships between soil, water and human cultures, focusing on the causes and effects of the disappearance of tropical rain forests. The 70-minute production, available to schools, festivals and regional theaters, is their most ambitious yet, and features over 100 vibrantly colored two- and three-dimensional images from 1 to 15 feet high. In *The Water Tree*, as in their earlier works, the company delivers an environmental message that not only delivers the bad news but tickles the funnybone and fires the imagination as well.

From *Up River/Down River*, January/February 1991. Reprinted by permission.

 EARTH CARE ACTION

He Speaks for Salmon

By Stephen Stuebner

In the summer of 1968, Ed Chaney remembers standing beside the Columbia River, marveling at thousands of salmon fighting the current below Oregon's John Day Dam. In The U.S. Army Corps of Engineers' rush to prepare the dam for its September christening, it had sealed off the river before the fish ladders were ready.

Adult salmon heading for spawning grounds hundreds of miles inland were blocked repeatedly by the monolithic structure and perished. Army Corps officials denied the problem, but Chaney had proof: candid photographs of dead fish washed up on shore. A full page of these photographs in the *Portland Oregonian* brought attention to the problem. "The fire alarms went off," says Chaney.

For the past quarter-century, Ed Chaney has battled to improve conditions for young salmon in the American Northwest. (Glenn Oakley)

A quarter-century later, Chaney, a six-foot, six-inch genteel Idaho conservationist, is still trying to save the salmon, only the situation has gotten much worse. Snake River sockeye have been designated as an endangered species, and three Snake River chinook runs have been listed as threatened or endangered. Other runs, such as the Snake River coho, have become extinct. The American Fisheries Society estimates another 150 salmon runs throughout the Northwest are in trouble.

In the Lower Snake and Columbia rivers, Chaney blames eight dams that block key migrating corridors for killing as much as 95 percent of the juvenile salmon each year. However, the Army Corps of Engineers and the Bonneville Power Administration (BPA), which manage the world's largest hydroelectric system, have yet to make significant changes in dam operations to save the fish. Meanwhile, electric utility officials lobby government authorities against change to protect the lowest electrical rates in the nation: two to four cents per kilowatt-hour.

Tireless Volunteer

For Chaney, a natural resources consultant, the task of saving the salmon has consumed his life. He frequently works 20 hours a day, sacrificing his personal life in the process. "If you're going to fight a battle," he says, "you might as well fight a big one, and they don't come any bigger than this. We're wiping out a resource that has enormous social, cultural and economic value."

I f you're going to fight a battle, you might as well fight a big one, and they don't come any bigger than this. We're wiping out a resource that has enormous social, cultural and economic value.

In the mid-1970s, when commercial and sport fishermen and Indian gillnetters were arguing over who was responsible for dwindling fish runs, Chaney voluntarily wrote the first comprehensive salmon status report for federal and state officials in the Columbia River Basin. It revealed with empirical evidence that dams—more than fishermen—were wiping out the fish. In 1980, he helped convince Congress to create a special agency charged with giving fish equal status with hydropower on the Columbia.

Though the BPA has spent $1 billion in the last decade on fish, Snake River salmon runs have continued to decline toward extinction. The Army Corps barges most of the juvenile fish around the dams, but few of the transported fish return to the Columbia as adults. In the wake of endangered species listings, authorities have called for reservoir drawdowns to give water back to the fish. But industrial interests are fighting the plan with predictions of economic ruin.

Chaney disputes such predictions. He and the 40 or so environmental and Indian groups in the Pacific Northwest that he helped bring together into a coalition have come up with an economically viable solution to save at least a portion of the once-magnificent salmon runs. The battle could last for years, though for the fish, time is running out.

"If the salmon ever recover to fishable numbers in our state," says Idaho Governor Cecil Andrus, "people should name a tributary of the Salmon River after Chaney. That would be the right recognition for a man who has relentlessly devoted 20 years to the well-being of anadromous fish."

From *National Wildlife*, October/November 1992. Reprinted by permission.

Heaven at Her Doorstep

By Mark Wexler

Ballona Lagoon is small compared to most tidal wetlands. The 16-acre, 100-foot-wide estuary, which runs through the tony section of Los Angeles known as Venice Beach, is one of the last remnants of a vast saltwater wetland system that once covered hundreds of acres along the southern California coast.

Despite its size, however, Ballona Lagoon provides valuable habitat and feeding grounds for more than 20 species of migratory waterfowl and shore birds, including the endangered California least tern. It also contains more than a dozen species of fish and clams, populations of several types of crabs and at least two dozen species of native plants. "It is," says Venice resident Iylene Weiss, "the most productive saltwater wetland within L.A. city limits."

When Weiss and her husband bought a house adjacent to the lagoon in the mid-1980s, it was like a dream come true for the former founder of the L.A. chapter of the Oceanic Society. "To have an estuary at my doorstep, that was heaven," says the mother of five grown sons.

But Weiss's dream turned into a nightmare when she learned of a developer's plan to dredge Ballona Lagoon into a boat marina. The self-proclaimed "full-time professional volunteer" decided she wasn't going to let that happen without a fight. It wasn't an easy task. First she had to convince neighbors that a wetland would be more valuable to them than a marina.

Then she had to take on a developer who had huge financial resources.

*T*o have an estuary at my doorstep, that was heaven. But the dream turned into a nightmare: a developer's plan to dredge Ballona Lagoon into a boat marina.

With other concerned citizens, she organized a grassroots campaign to teach the community about the ecological significance of Ballona Lagoon. She also devised a clever plan to take advantage of the developer's resources.

"To apply for a building permit with the California Coastal Commission," says Weiss, "the developer had to provide an environmental impact statement. We also filed for a permit, but our plan called for restoring the lagoon. Then we just sat back and waited for the developer to finish his study and inadvertently provide us with the scientific data we needed to make our point."

Victory!

Though Weiss and the other members of her group, the Ballona Lagoon Watch Society, had no experience with federal or state regula-

Thanks to the efforts of Iylene Weiss and her fellow volunteers, the banks of Ballona Lagoon in Los Angeles are again alive with a variety of flowering native plants. (Dave Butow, Black Star)

tions, they successfully defeated the marina plan. "Iylene took the initiative to gain protection for one of our city's most overlooked natural treasures, at a time when other people were seeking to destroy it," says L.A. City Councilwoman Ruth Galanter.

The group's victory was still incomplete, however. The degraded lagoon needed to be restored to healthy condition. To help do so, Weiss's group convinced the Coastal Commission and the Environmental Protection Agency to provide more than $100,000 in grants to develop a wetland restoration program.

That program has now begun. Today, the banks of Ballona Lagoon are again alive with a variety of flowering native plants. Work also is under way to improve the tidal circulation that flushes pollutants from the estuary. And Weiss, who has received national honors for her efforts, has turned her attentions to trying to save other remaining southern California wetlands.

From *National Wildlife*, October/November 1992. Reprinted by permission.

Long Island Soundkeeper

By Elaine Pofeldt

As a third-generation fisherman, Terry Backer has a personal stake in his work as Long Island Soundkeeper. "To me, Long Island Sound is potatoes," Backer recently told a high school class. "It's where my family obtained its sustenance." But as an estuary, the sound is also a critical nursery for the plants and creatures that populate the Atlantic Ocean. It is one, he says, that has for too long seen industrial discharge of heavy metals, inadequately treated sewage and polluting runoff from the land. In recent years, its harbors have seen massive fish kills from hypoxia, a lack of sufficient oxygen in the water caused by excess bacteria.

That doesn't surprise Backer. About 10 percent of the U.S. population lives within the sound's drainage basin, which includes parts of New York, Connecticut, Rhode Island, Vermont, Massachusetts and New Hampshire. "The communities we live in are very shortsighted, particularly the ones that are in financial trouble." he says.

"But in some ways, the sound is better than it was 30 years ago," says Backer. "The load of heavy metals in its waters has decreased with the decline in smokestack industries on its shores." Backer fights other pollution sources from his office in Norwalk, Connecticut, and from the boat in which he patrols the sound for pollution. He is well known to legislators for his persistent lobbying on environmental protection laws, and to in-

dustry representatives for his tough negotiating style. "The real battle for oceans begins in creeks and estuaries," he says. "There's still a sense among the population that these places are swamps and mosquito holes. Well, they're also lobster holes. If we lose the mudfight in the estuaries, we're also going to lose the larger species."

> *T*he real battle for oceans begins in creeks and estuaries. There's still a sense among the population that these places are swamps and mosquito holes.

Backer's mudfighting recently led to protection for the sound's shellfish from a planned natural gas pipeline. The Iroquois Gas Transmission Systems Company originally proposed locating the line directly on the sound's floor. Arguing that this would block migration by lobsters and other bottom-dwellers, Backer convinced the company to bury part of the pipeline, though it would cost them millions. "I came to an agreement in an hour by going in and being honest with them," Backer says.

Backer's sincerity has been a powerful asset in his efforts to enact changes that protect the sound. "I think what I bring to it is the

Terry Backer—no landlubber he—has a high personal stake in the quality of the local waters. (Courtesy of Long Island Soundkeeper Fund, Inc.)

average person," says Backer, 37. "I'm your average Joe Shmoe who cares about the place I live in. And I've always been one of those 120 percent people, the kind that makes you nauseous because it's 10 P.M. and you want to go home."

Softening Resistance

Backer began acting on his passions after dropping out of high school to seek adventure as a logger and fisherman in Alaska, and as a traveler in South America, Europe and Africa. "I came back to work with my father on a

lobster boat," he recalls. "I said, 'Holy mackerel, what's all this crap in the water?' He said, 'That's what it is.' I said, 'This place has gone downhill since I left.' He said, 'You should see it from where I'm standing.' " Backer's father had lived and worked on the sound for 60 years.

After helping establish the Connecticut Coastal Fishermen's Association, Backer eventually spearheaded their suit against several municipalities that had violated their sewage discharge permits. Part of the out-of-court settlement was used to establish the Soundkeeper Fund in 1987, with Backer at the helm. "One day I'm going to have to answer to my children for where I'm standing," says Backer, who has two young sons. "Not doing something would have been arrogance on my part."

One of Backer's most pressing cares is raising enough public awareness of the sound to ensure the better land use planning, environmental law enforcement and individual behavior changes needed to reduce the sound's pollution. "You can't just throw money at a problem," Backer says. "A lot of the changes don't cost money. It's a matter of breaking habits." Yet Backer stresses that environmental agencies need more funds to enforce existing laws. "People don't want to make commitments when they're not perceived to be materially satisfying," Backer says, "My job is to soften their resistance to change."

Reprinted with permission from *E—the Environmental Magazine,* July/August 1991. Subscriptions $20/year; P.O. Box 6667, Syracuse, NY 13217; (800) 825-0061.

Fixing a Broken River

By Leah Barash

From the start, it had all the makings of an ecological nightmare. In 1964, the U.S. Army Corps of Engineers began a questionable project to build a 110-mile-long shipping canal across Central Florida, linking up the Gulf of Mexico with the St. Johns River and Jacksonville. If completed, the proposed Cross Florida Barge Canal would have destroyed tens of thousands of acres of wildlife habitat and contaminated the aquifer that supplies drinking water to the region.

As it turned out, by 1969, the Army Corps had completed a third of the project and in the process had flooded some 9,000 acres of riverine forest to create the massive water-storage dam and lake called Rodman Reservoir. The dam blocked the migratory routes of such aquatic species as manatees, striped bass and shad. Black bears, river otters and other riparian wildlife also found themselves cut off from feeding corridors. And when the project destroyed some 16 miles of the swift-flowing Oklawaha River and its vast flood-plain, Marjorie Carr could no longer sit back silently.

The energetic mother of five, who has a master's degree in zoology, had cherished the Oklawaha's beautiful serpentine course ever since she first laid eyes on the river in the 1930s. In 1969, she rallied other conservationists to form the Florida Defenders of the Environment (FDE), based in the Gainesville home she shared with her husband, the late Univer-

Jack Kaufmann and Marjorie Carr are dam angry about the blockage of the Oklawaha River. (Robert Holland)

sity of Florida zoologist Archie Carr. The group helped convince authorities to stop the canal's construction.

Reversing Disaster

Though the project may have died, Rodman Reservoir continues to haunt the Sunshine State. Because the nutrient-rich waters of the Oklawaha are trapped behind the dam, they cause a buildup of noxious weeds and algae in the reservoir.

"The reservoir now has to be drawn down periodically to kill the weeds, to the tune of more than $1 million annually," says Jack Kaufmann, a University of Florida zoologist who joined forces with Carr and the FDE in 1970 to develop a plan for taking down the dam and restoring the Oklawaha to its former meandering course—a plan that ultimately, according to a University of Florida economist, would save taxpayers enormous amounts of money.

Though Florida officials recommended as far back as 1976 that the river be restored, the FDE's efforts continue to be stymied by central Florida political factions that favor leaving the reservoir intact. The reason: Rodman's reputation as a prime bass fishing spot brings needed revenue to the region. "Opponents of river restoration assume that if we drain Rodman, the fishermen won't spend any more money," says Kaufmann, who has devoted thousands of hours to the battle. "The return of the river would actually increase recreational opportunities not only for fishermen but also for other outdoor enthusiasts."

Currently, Carr and Kaufmann are serving on advisory committees to help the state solve the dispute. They're also busy trying to rally support for their plan. "People say, that's controversial, I'll stay out of it," notes the feisty, 78-year-old Carr. "Well, hell, there's nothing controversial about this—the facts are so overwhelmingly in favor of restoration." It's tough to argue with the woman who has devoted three decades of her life to reversing an ecological disaster.

From *National Wildlife*, October/November 1992. Reprinted by permission.

When Pollution Hits Home

By Michael Lipske

Lower Beaverdam Creek and its junked cars are not one of Washington, D.C.'s major tourist attractions. But it's just the sort of place local environmental activist Norris McDonald takes people when he wants to show why the city's Anacostia River is such a mess.

The narrow brown creek wends past automobile chassis, rusted fenders, mufflers and tires before joining the Anacostia, a sluggish tidal river that the *New York Times* labeled the national capital's "backyard refuse pit."

Stripped of wetlands, choked with silt from suburban development, laden with pollution from street and parking lot runoff and from sewage overflow during storms, the Anacostia is Washington's "other river." The better-known Potomac River has been scrubbed with a multimillion-dollar cleanup campaign. But the dirty Anacostia "has been literally trashed to the point where, in some areas, it is unfit for human contact," says D.C. delegate to Congress Eleanor Holmes Norton.

Norris McDonald hopes to change that situation. Four years ago, he founded the Center for Environment, Commerce and Energy, one of the nation's few organizations devoted to solving environmental problems of African-Americans. One of his immediate goals has been to focus attention on the Anacostia and pollution problems in the low-income neighborhoods along its banks.

The Value of Water

McDonald organizes Anacostia cleanup days, leads creek walks, conducts water-quality tests and persuades businesses to "adopt" stretches of the fouled suburban streams that flow into the river. But his real task, he says, "is mobilizing the African-American community to aggressively participate in our environmental future."

To better demonstrate to District of Columbia residents the links between the Anacostia River and themselves, McDonald came up with his WET (Water, Education and Training) program, which identifies and fixes leaks in toilets, sinks and bathtubs in homes and apartment complexes. "Just getting people aware of the value of water as a resource," he says, can do more than trim the family utility bill.

Stopping leaks, McDonald maintains, also reduces the flow of wastewater to the city's sewage treatment plant, a major electricity user. Less water to treat means less use of power, reducing the need to build more electrical generating facilities in the city, thereby curbing potential air pollution. "Through our WET program we hope to educate people to the bigger environmental picture," says the 39-year-old activist.

"Norris has been ahead of the curve for a long time," says David Hahn-Baker, a Buffalo,

Norris McDonald pulls his daily catch from the Anacostia River. (Robert Rathe)

New York, environmental consultant to the National Wildlife Federation and other groups who specializes in urban-related issues. "He's been a leader in addressing nontraditional environmental problems in the context of people-of-color communities."

McDonald says that his dream is to turn his organization into a powerful voice for all African-Americans fighting for environmental fairness, as well as a source of solutions to conservation problems. By focusing on Ana-

costia River pollution and other dilemmas that hit close to home, the D.C. activist believes his most important environmental message is self-sufficiency. "If somebody else dirties up your yard, yes, scream at them and try to get them to clean it up," says McDonald. "But if all else fails, be prepared to clean it up yourself."

From *National Wildlife*, October/November 1992. Reprinted by permission.

Seeking Sound Evidence

By Jeann Linsley

With more than 2,700 miles of shoreline to monitor, Puget Sound would represent a difficult challenge for even the most well-staffed pollution control agency. For Washington's beleaguered Department of Ecology, which is charged with enforcing state and federal water quality laws, the task is all but impossible. "We're strapped for people and money," admits Kevin Fitzpatrick, acting head of the department's industrial permit section. The result: a lack of enforcement efforts and increasing levels of contamination in the 2,200-square-mile sound.

Eight years ago, to help deal with the situation, a number of concerned citizens joined forces with some Washington environmental groups to form the Puget Sound Alliance. The watchdog coalition began looking for innovative ways to use citizen volunteers to monitor water contamination. It found Ken Moser.

A former sailing ship skipper with a keen interest in environmental law, Moser was selected by alliance leaders in 1990 to become the official Puget Soundkeeper. As such, he heads up a citizen pollution-monitoring program similar to other "keeper" efforts already operating in San Francisco Bay, Long Island Sound, and the Hudson and Delaware rivers. Under Moser's leadership, nearly 200 volunteers now regularly patrol the banks and bays of Puget Sound in kayaks, small boats and on foot, looking for evidence of illegal contamination.

Nearly 200 volunteers now regularly patrol the banks and bays of Puget Sound in kayaks, small boats and on foot, looking for evidence of illegal contamination.

"These citizens are trying to fill a vacuum that our legislators are not allowing our agencies to fill by providing adequate funding," says Vim Wright, an alliance founder and assistant director of the University of Washington's Institute for Environmental Studies.

The volunteers who join Moser attend workshops where they study everything from Puget Sound plankton to details of the federal Clean Water Act. They also tour sewage treatment facilities to learn what kinds of wastes to look for outdoors. The actual sampling of water is left to Moser, who is specially trained in handling hazardous substances, and to government experts who assist the soundkeeper program.

"Each case takes a tremendous amount of surveillance and research time," says Moser, who frequently must weed through stacks of

Skipper Ken Moser (with beard and cap, left) leads a volunteer army in patrols of Puget Sound looking for eco-bad guys. (Phil Schofield)

state and federal records to determine if a potential violator has a history of noncompliance with the law. "Our first option," he adds "is to try to give violators the chance to correct problems. But if they resist and continue to pollute, then we have no choice but to file a citizens' lawsuit to stop them."

To date, the pollution fighters have successfully stopped the flow of contamination from at least a dozen industrial sites; their activities have also cost polluters more than $150,000 in fines. And currently, the group is challenging the legality of state discharge permits for 12 paper and pulp mills in the region.

"We've accumulated data to show that the permits are not adequately stopping the mills from releasing thousands of gallons of dioxin daily into state waterways," says Moser, whose only regret is that he cannot spend much time these days on the water. "There's so much work to do on land, nailing down all of the evidence, that my kayak may start getting a little rusty."

From *National Wildlife,* October/November 1992. Reprinted by permission.

Much of America's wilderness has been lost to highways, shopping malls and housing developments, but the majestic beauty that remains is nowhere more evident than in Arizona's Grand Canyon. (Carl Doney)

WHOSE WOODS ARE THESE?

By Michael D. Lemonick

Deep inside the dwindling woods, a rare species of bird is threatened with extinction. Before loggers came to the forest, the birds could easily find the trees they needed for nesting—trees at least 80 to 100 years old. But the relentless advance of chain saws has leveled much of the old woodland. Environmentalists filed suit under the Endangered Species Act and won, forcing the government to put new restrictions on logging in portions of several national forests. No one knows, however, if the action came soon enough to save the endangered birds—or the unique habitat that is their only home.

This story may sound familiar, but these forests are not in the Pacific Northwest, and the bird in question is not the northern spotted owl. It is the red-cockaded woodpecker, a striking red-black-and-white bird that lives in loblolly and longleaf pines from Virginia to Texas. Like the owl, it is being used by biologists as an indicator species, a sensitive probe of the vitality of forests across a broad swath of the United States. Just as dying canaries

once let coal miners know that oxygen levels were perilously low, the decline of the red-cockaded woodpecker, the northern spotted owl and many other species is a warning of a far greater threat: America's few remaining stands of old-growth forests—woods whose ancient trees have never been logged—are in danger of disappearing as distinct and valuable ecosystems.

That danger has sparked increasingly bitter court battles and legislative debates between the forces of conservation and the defenders of the logging industry. In addition to campaigning on behalf of the woodpecker and the owl, environmentalists have demanded protection for the northern goshawk in Arizona and New Mexico and the grizzly bear in the northern Rockies. In Alaska protesters have forced the government to reexamine timber sales from the wild Tongass National Forest.

Many conservationists call for nothing less than an overhaul of U.S. forest policy—a policy, they charge, that too often treats the woodlands as resources to be ex-

213

The still-pristine woods along the Kadashan Bay in Alaska's Tongass National Forest are scheduled for future harvests. (Mark Kelley/Alaska Stock Images)

ploited rather than heirlooms to be preserved. Various bills being considered in Congress range from a proposed reduction in logging in national forests to more modest measures that would at least stop the government from selling timber at a loss, a practice that in effect charges taxpayers money to have publicly owned forests destroyed.

Pressure for change is also coming from within the U.S. Forest Service, the agency responsible for the national forests. The supervisor of Montana's Lolo National Forest recently halved the amount of logging he would allow, citing the threat to native elk and the danger of soil erosion. Dissatisfaction has led some 2,000 Forest Service employees to join the two-year-old Association of Forest Service Employees for Environmental Ethics. Says founder Jeff DeBonis, a former timber-sale planner for Oregon's Willamette National Forest: "The problem is obviously a lot bigger than just owls."

Enchanted Timber

More than 95 percent of the virgin forest in the lower 48 states has already been lost to agriculture, development and logging, and Alaska's woods are disappearing as well. At least 75 percent of what remains lies in 156 national forests. These tracts embrace large expanses of pristine wilderness, a haven for nature lovers and also for more than 3,000 kinds of vertebrates and thousands of species of insects and plants.

But since its creation by Theodore Roosevelt in 1905, the Forest Service has been required to manage its lands for "multiple use." The national forests now hold some 18 percent of the country's commercial timber, serve as cheap grazing land for thousands of cattle, support multi-million-dollar mining operations and contain a 579,000-kilometer (360,000-mile) network of roads that is eight times longer than all the U.S. interstate highways combined.

*T*he problem is obviously a lot bigger than just owls.

Until World War II, the national forests comfortably accommodated all users, but that was when these lands were being called on to provide less than 5 percent of the nation's wood. Then came the postwar building boom, which boosted demand: The annual cut from the national forests surged from 3.5 billion board feet in 1950 to 9.4 billion board feet a decade later. (A board foot, the standard unit used to measure timber, is equivalent to a slab of wood 1 inch thick and 12 inches square.) This dramatic growth worried some members of Congress, which passed laws requiring the Forest Service to balance the goals of conservation and resource development.

But the outflow of timber from the national forests still amounts to about 10 billion board feet a year, 14 percent of the nation's output. One reason is that the biggest, oldest trees, which contain enormous

quantities of high-quality wood, have all but disappeared from private lands. Moreover, the lumber from the national forests is cheap for logging companies. The Forest Service not only builds all the roads, providing easy access for workers and machinery, but also sells off the trees for prices that are often far below market value. Perhaps most significant, the decisions on how much timber will be sold each year come, in the end, from Congress. Big lumber companies have enormous political clout in heavily forested states, and the Forest Service is considered by environmental groups to be little more than a federally subsidized logging agency. Says David Wilcove, an ecologist at the Environmental Defense Fund in Washington: "Wildlife and wildlife habitat on federal lands are being sacrificed for the sake of the timber quota."

*W*ildlife and wildlife habitat on federal lands are being sacrificed for the sake of the timber quota.

Within the Forest Service, however, there is now a rising tide of antilogging sentiment. During the 1970s and 1980s, the agency recruited ecologists and wildlife-and-fisheries biologists, in addition to its traditional foresters, road engineers and timber managers. While Forest Service policy didn't change much, the average level of environmental consciousness did.

Last September [1990], John Mumma, a regional forest manager based in Missoula, Montana, told a congressional subcommittee that he had been transferred from his position when he refused to meet timbercutting targets. Testified Mumma: "I have failed to reach the quotas because to do so would have required me to violate federal environmental law." He later resigned. More recently Ernie Nunn, supervisor of the state's Helena National Forest, failed to meet his forest's prescribed cut and was told by Mumma's successor to plan on a reassignment.

But on the 800,000-hectare (two-million-acre) Lolo National Forest in western Montana, another manager successfully resisted prescribed cutting. Lolo supervisor Orville Daniels decided that meeting his assigned quota of timber sales would deprive elk herds of critical escape cover and also pose silt runoff problems in the highly erodible granitic soils. So he put the sales on hold for ten years and beat back a timber industry group's attempt to have the decision reversed; the area remains off-limits to loggers.

The God Squad

Environmentalists have learned to regard such victories with a wary eye, especially in the wake of the spotted owl debate. That controversy seemed to be settled in April 1990 when a government task force determined that the bird was imperiled and that about 1.5 million hectares (3.8 million acres) of its old-growth habitat on federal forest land should be protected. The U.S. Fish and Wildlife Service later proposed logging restrictions on 4.7 million hectares (11.6 million acres) of owl habitat. After a year and a half, the Bush administration has yet to agree to a plan.

Until it does, a federal judge ruled last May, sales of timber on 26,700 hectares

(66,000 acres) of national forests in the Northwest must be suspended. In the meantime the administration, whose words are pro-environment but whose actions often are not, is considering a way to maintain logging on other publicly controlled lands. In October Interior Secretary Manuel Lujan Jr. announced he would convene the so-called God Squad, a Cabinet-level committee that can override the Endangered Species Act in the regional or national interest. The squad will decide by mid-March whether to permit logging on some 1,850 hectares (4,570 acres) of Oregon land administered by the Bureau of Land Management rather than the Forest Service. The committee has convened only twice before—in its best-known action, it refused to allow the Tennessee Valley Authority to build a dam because of a threat to a small fish known as the snail darter. Says Michael Bean, senior attorney at the Environmental Defense Fund: "If the God Squad grants an exception [to the Endangered Species Act], it's clear what will happened in subsequent cases."

For Love of Money

Environmentalists have suffered a setback in another case: last August a federal judge refused to issue a restraining order to stop the cutting of old growth ponderosa pines in six national forests in Arizona and New Mexico. Environmentalists claimed the logging was endangering local populations of the northern goshawk, a predatory bird being touted as the spotted owl of the Southwest. The Forest Service is expected to release its final guidelines regarding logging in the birds' habitat early next year [1992].

The logging industry contends that excessive concern about birds could carry high costs for humans. In the Pacific Northwest, the companies say, nearly 100,000 jobs (a figure disputed by environmentalists) would be lost if the northern spotted owl were protected. Industry officials deny that all the old-growth forests could ever be wiped out, since some of these woodlands are already barred or inaccessible to loggers. Moreover, most companies claim that they plant more trees than they cut.

The number of jobs in logging and related industries has plummeted in recent years, largely because of automation. If logging were cut back on federal land, much of the slack could be taken up by privately owned and less environmentally sensitive woods. Besides, small communities that really depend on government-owned woods for jobs will run into trouble as the forests are stripped away; not only will the trees be gone but also land that might otherwise be touted for recreation will become profoundly ugly.

The Forest Service logging operations are questionable on economic grounds as well. While the agency claims it made $628 million in profit last year, critics dismiss that figure as absurd. Robert Wolf, a forestry expert and emeritus economist with the Congressional Research Service, who recently analyzed the Forest Service accounts, told the House Agriculture Committee in October that "these mythical profits are achieved by accounting alchemy." Among the creative bookkeeping methods Wolf found were a failure to subtract $327 million that the agency paid to states in lieu of taxes and a practice of spreading the costs of building logging roads over hundreds of years. Wolf says

that even with a $700 million congressional appropriation, the Forest Service timber program had a negative cash flow of $186.4 million in fiscal year 1990—and similar losses for a least the past four years.

Some individual forests do make money, but they are more than offset by places such as the Tongass National Forest in southeastern Alaska. More than three times the size of Massachusetts, the Tongass is the largest remaining temperate rain forest in the United States and still contains about two million hectares (five million acres) of old-growth trees. Until last year, two 50-year contracts required the Forest Service to sell off 4.5 billion board feet of Tongass timber every decade—even when the price of lumber fell in the early 1980s to a rock bottom $1.50 per 1,000 board feet.

> *The national forests by definition belong to the American public, and it is the public, not industry lobbyists or agency bureaucrats, that should decide their fate.*

After years of campaigning by activists, Congress last year passed the Tongass Timber Reform Act, which closed off to loggers an additional 400,000 hectares (one million acres) of the forest, nearly 25 percent of it old growth. The bill also called for modifications in both of the long-term logging contracts, raising the official price of some types of Tongass timber to as much as $568 per 1,000 board feet. The federally mandated logging quota was replaced by provisions that the amount of timber offered for sale should equal market demand. But critics of the Forest Service say the agency is largely ignoring the new laws and preparing to sell Tongass timber faster than ever before.

Forests for the People

Environmentalists believe much broader legislation is needed. Their favorite proposal is the Ancient Forest Protection Act of 1991. It calls for the government to identify ecologically significant old-growth forests across the nation and protect these areas from environmentally disruptive exploitation, including logging. Another bill would ban money-losing federal timber sales and force the Forest Service to calculate the profitability of these operations in a more realistic way. Says Representative Jim Jontz, an Indiana Democrat who is principal House sponsor of both bills:

"The Congress has to decide. Are we going to discard environmental laws and values, or are we going to bring down the level of timber production?"

The national forests by definition belong to the American public, and it is the public, not industry lobbyists or agency bureaucrats, that should decide their fate. With increasingly concerned managers inside the Forest Service, environmentalists on the outside and legislators looking sharply over the agency's shoulder, there is reason to hope that the last stands of ancient trees will remain uncut—and that some of their younger cousins will eventually achieve the status of old growth themselves.

The Forest-Friendly House

By Wilbur Wood

The house of the future—or at least one version of it—nears completion in Missoula, Montana. Though it appears "mainstream" enough, this house is constructed, inside and out, almost totally from recycled or reused materials. The designer and builder, Steve Loken, calls it a "national demonstration for innovative building technology."

Missoula lies in the heart of timber country, but excessive cutting of the slow-to-regenerate old-growth forests in the northern Rockies (much of it for export to Japan) has meant that local builders rarely find quality lumber at affordable prices anymore. So Loken has sought substitutes, and his "Recraft '90" house, as he calls it, will contain almost no virgin lumber.

What it will contain—from foundation and frame to siding and roofing, from walls and floor panels to carpets and tiles—is wood waste, cardboard, newspapers, straw, rubber tires, plastics, mining slag and glass. Outdoor paving bricks will be formed out of compacted sewage sludge, carpets will be made from recycled wool and plastic resins and floor tiles from recycled auto windshields. Even the nails will be second-hand or remanufactured.

Built to Last

Loken has a twin agenda: to help save some of the last remaining old-growth forests (the United States uses 20 percent of the world's lumber from such forests, yet only 3 percent of the world's old growth remains here) and to encourage manufacturers to retool and create high-quality building products out of "trash" timber, wood wastes and other fiber wastes.

"A new generation of environmentally responsible companies needs to start reprocessing this stuff," he says. Prices for currently available waste-based building materials are, he adds, "in line with everything else we're building with"—not expensive, not cheap. "But the cheapest doesn't last," he says, "and it's time to build things that last."

Loken intends to sell the Recraft '90 house for around $120,000. First, though, he and his family may live in it awhile to monitor performance of the unconventional materials and to test the energy- and water-saving technologies embodied in its passive-solar-heated, superinsulated design. They should have quite

---

WHAT YOU CAN DO

When camping or hiking through the woods, carry out *everything* you brought in—including tinfoil, bottle tops, cigarette butts and sanitary napkins.

219

a housewarming party. (Contact: Steve Loken, Center for Resourceful Building Technologies, P.O. Box 3413, Missoula, MT 59806, 406-549-7678.)

Reprinted with permission from *E—the Environmental Magazine*, November/December 1991. Subscriptions $20/year; P.O. Box 6667, Syracuse, NY 13217; (800) 825-0061.

 EARTH CARE ACTION

Small Is Beautiful

By Marion Lightly

In 1936 Canadian Merv Wilkinson bought 55 hectares (136 acres) of land at Yellow Point, a peninsula just south of Nanaimo, British Columbia, on Vancouver Island. Hoping to turn it into a farm, he went off to the University of British Columbia to study agriculture. It was there, through the influence of one of his professors (who happened to be a forester trained in Sweden), that Wilkinson decided to learn forestry instead. After graduation, he named his property Wildwood and began to manage the land as a woodlot.

Wildwood is a perfect example of the dry coastal Douglas fir forest. Virtually all species of trees indigenous to eastern Vancouver Island are thriving there; there are even some Douglas firs that Wilkinson figures are over 1,800 years old. The climate is moderate, the terrain is gentle (making roads easy to build), and it is close to a lumber market—a forester's dream.

Wilkinson's dream was to manage a coastal forest that would produce wood products without losing all the characteristics he had come to love and respect. His vision must have been a clear one because today he has a healthy, thriving woodlot that has provided him with about a third of his income over the last 50 years. And despite cutting approximately 1.4 million board feet during that time, Wilkinson figures that—prior to making his ninth cut in 1990—he has the same amount of timber standing as he did when he began in 1938.

Working with Nature

Wildwood is living proof that you can have your forest and mill it too—at least on the small scale. Its name is especially appropriate; the property resembles a wilderness forest, even though human management is always giving nature a gentle push to produce high-quality wood products. Wilkinson estimates that he has increased the growth rate of his trees by an average of 12 percent.

The forester attributes his success to "common sense" and "working with nature." One technique he uses employs the basic principles of genetics. Trees with desirable characteristics—tall, straight trunks, abundant foliage, and good cone production—are left throughout Wildwood to provide seed for future generations. This is the best way to ensure that there will be a steady supply of healthy young trees that are genetically adapted to the site they germinate in. While other practitioners of "selective logging" often send these prime trees to the mill and tally up the profit right away, Wilkinson sees a larger (albeit long-term) profit in leaving the parent trees to do what they do best—maintain the quality of his forest in perpetuity.

*F*oresters get little chance to have a long-term relationship with the area they are working in because they are moved around so much.

This approach to working with nature takes patience, and Wilkinson has spent a lot of time observing the trees on his woodlot. For example, he has recognized that in certain parts of his forest the cedars and firs grow in cycles. "When I first came here there were cedars. I took them out and they were replaced by firs. Now I'm cutting the firs and the cedars are coming again. This is nature's way of ensuring against disease and soil depletion," he says. He points out that in this country, foresters get little chance to have a long-term relationship

with the area they are working in because they are moved around so much. "In Europe," he adds, "they have more continuity—some of them stay with the same forest for 40 years or more."

Relying on Woodpeckers

Wilkinson also uses animals as indicators of conditions in the forest. While a lot of foresters will take out a tree as soon as it's dead because they feel it is a hazard, he relies on woodpeckers to tell him when it's time to take out an old snag. "You can see them move their nests down each year as the snag rots. When they abandon the tree altogether, it's time to take it down—it's not safe anymore." In the meantime, he says old snags can be very beneficial; they provide homes for insects, birds, spiders and other species that maintain the balance of life in the forest and prevent outbreaks of destructive insects.

The British Columbia woodlot owner doesn't do any clear-cutting either, nor does he burn slash or use chemicals to remove undergrowth. The canopy is left intact in order to keep the soil shaded and moist, but he opens it up through thinning to let in enough light to encourage the growth of young trees. Limbs and debris are scattered about on the ground to replenish the soil, and he practices some brush control to cut down on competition with the growing seedlings.

There is a delicate balance here, he says, because in some ways the brush is beneficial. It protects the seedlings, harbors birds and wildlife and can replenish nutrients in the soil. But it can also be a hindrance when it competes with the very young seedlings for light, water and nutrients. He has learned through the years that sheep, given supplemental feed and limited access, are very effective brush control agents. Unlike deer, they don't seem to like eating the young conifer tips. But Wilkinson only lets them graze while the seedlings are small—once the trees are free growing, he removes the animals and leaves the brush alone.

Of course, the forester and other practitioners of selective logging are not without critics. Clear-cutting advocates say such methods don't open up the forest enough to allow healthy regeneration of young trees. In response, Wilkinson points out that trees grown in the open have trunks with a lot of thick lower branches, widely spaced growth rings and a conical shape—all of which produce poor-quality lumber, if they're used for lumber at all. Most end up as pulp. In contrast, saplings grown among taller trees have to stretch for the light in their early years and end up with tall, straight trunks and fewer and thinner lower branches. This leads to tight-grained, knot-free wood with a much higher market value.

Two other arguments against Wilkinson's methods relate to sustainability. One is that selective logging that uses skidders to haul out the logs (as his does) will result in soil compaction and damage to roots after five or six passes. This, of course, only applies to the trees whose roots are under skid trails that are used over and over. Although evidence of such damage is not immediately visible, it could show up as die-back at some point in the future.

The other major concern is root rot. This is a condition caused by fungi inside the root tissues—the two common species on eastern Vancouver Island are *Phellinus weirii* and *Armillaria mellea*. While it is common for the roots of apparently healthy trees to be infected, the fungi are normally in a suppressed condition. However, when a tree is cut down, the fungi spread to neighboring trees, wherever there is root contact.

A Model to Follow

While "antiselective" foresters warn of potential outbreaks of root rot in Wilkinson's forest, he is probably safe. Each species of fungus has a preferred tree host, so where there is a mixture of tree species, as is the case at Wildwood, the spread of this condition is kept in check. Also, even in clear-cut operations, the roots are left in the ground, where the fungi responsible for this condition may remain infectious for decades.

The question now being debated in British Columbia is whether selective management can be successful on a large-scale commercial basis. David Handley, a forester for MacMillan Bloedel, has prepared papers on why selective logging would not work. While the arguments sound convincing, it is easy to see that the economic issue overshadows any biological or ecological considerations. Selective logging is more costly in the short term, especially road-building, and profits are a lot further down the path.

Even though Wilkinson has been managing Wildwood for 50 years with very promising results, it's still too early to tell if his initial success will continue indefinitely. Perhaps in another 100 to 200 years we'll know the answer. But in the meantime, he is providing a practical model of forestry that is a real inspiration to those who despair at our vanishing forests.

From *Nature Canada*, Winter 1992. Copyright © Marion Lightly. Reprinted by permission.

RESOURCES

As this shopper knows, shopping for environmentally benign products can be tricky. "Going for the Green," beginning on the opposite page, gives you a few valuable hints. (Francis Roberts)

GOING FOR THE GREEN

By Penelope Wang

So you want to be an environmentally correct consumer? Very commendable. But before you head to the supermarket, try this shopping quiz:

Which type of product is best for the environment? The one that is:

(a) recyclable,
(b) degradable, or
(c) ozone-friendly?

The correct answer is: (d) hard to say. None of the three terms tells you enough to be meaningful. Nevertheless, these days you are forced to decipher "green" claims on everything from toilet paper to refrigerators. Nearly 400 supermarket products launched last year [1991] boasted on their packaging that they were good for the environment, more than twice the number in 1988, according to Marketing Intelligence Service, a Naples, New York, research firm. Trendy retailers are pushing the green too. The Sharper Image now sells $99.95 automated can crushers for recycling, and Mo Siegel, founder of Celestial Seasonings, is starting Earth Wise, a manufacturer of household cleaners and trash bags. There are dozens of Earth-conscious mail-order catalogs with names like Ecco Bella and WeCare. And all over the country so-called eco-preneurs are opening environmentally friendly specialty stores.

Beware of the Greenwash

Unfortunately for consumers, some environmental marketing is mere greenwashing. "I would estimate that about 15 percent of the claims are true, about 15 percent are false, and the rest fall into a gray area," says Joel Makower, coauthor of *The Green Consumer*. The reason for the muddle is that few environmental terms on products have strict legal definitions. Complains Texas assistant attorney general Steve Gardner, "The attitude of marketers seems to be: If you can't make it better, make it up." And frequently, make it more expensive. "In a few cases, businesses are simply price gouging," notes Carl Frankel, editor of the *Green Market Alert* newsletter. "But more often, prices for green products are higher than competitors' because their manufacturers are too small to achieve economies of scale." The handmade Sun Frost refrigerator, for example, which uses less than half as much electricity as a comparable General Electric model, costs $2,350 versus $600 for the GE. At an annual

225

electric-bill saving of $60 a year, it would take more than 25 years to break even.

Some relief may be on the way. The Federal Trade Commission is now considering whether to issue industry guidelines on environmental labeling. But the agency isn't likely to act before year-end [1991], and even if it recommends changes, manufacturers will probably take several more months to implement them.

Safe, Safer, Safest

In the meantime, what's a well-meaning consumer to do? First of all, be skeptical of what appear to be seals of approval to help you hack your way through the green jungle. The not-for-profit group Green Cross now plants its certification insignia on 350 household products. A green cross means the organization has collected as much as $10,000 from a company to run tests that verify the manufacturer's green claims. Still, the insignia doesn't mean the product is totally safe for the environment. Green Cross gave its approval to Dolco's egg containers solely because they contained recycled polystyrene, even though the plastic foam does not disintegrate. A Green Cross spokesperson says that her group rates only specific claims, rather than reviewing a product's larger environmental effects.

To find the merchandise that is better for the environment and to identify those products with labels that are particularly vague or misleading, *Money* interviewed environmentalists, scientists, retailers and government regulators. They offered six shopping tips.

1. Choose products with less packaging than the competition. Each year the average American discards nearly a ton of trash, more than 30 percent of which is packaging. By looking for the least-packaged brands, you'll ease the strain on landfills and often save money, too. On a recent grocery-shopping trip to a Washington, D.C., Safeway, for example, Nissin Food's Cup O'Noodles soup (with vegetables and shrimp), packaged with a polystyrene cup, cost 79 cents for 2.25 ounces, while the company's Oodles of Noodles, without a cup (or the additional ingredients), cost 39 cents for 3 ounces—more than 50 percent less per ounce.

Keep an eye out for reusable containers. Downy fabric softener is now available in small cartons of concentrate (about $3.30) that you can use for refilling larger plastic bottles.

2. Try to buy products made of natural ingredients. Untreated materials generate fewer toxic chemicals in their production than processed ones. Companies that bleach paper white, for example, use a process that creates dioxin, a potential carcinogen. Household cleaners are also a major source of toxins. Some laundry detergents contain fluosilicate, often used as a pesticide. Annie Berthold-Bone, author of *Clean and Green* (Ceres Press), suggests using a combination of soap flakes and either Borax or Arm & Hammer washing soda, made with natural ingredients.

3. On labels, look for specifics about recycling, not merely the words recycled or recyclable. True, using products made of recycled material is generally better for the environment than using ones that aren't. But many recycled goods may be less helpful than you think. Manufacturers often label paper products and cans as recycled even if only some of their contents are reused material. For instance, only 25 percent of the average recycled plastic bottle is actually recycled.

Moreover, advertising a product as recyclable is meaningless if you can't recycle the material in your town. For example, only a handful of municipalities today recycle juice-box packaging, which contains compressed layers of paper, plastic and aluminum. In December, the New York City Consumer Affairs Department charged drink-box makers Combibloc and Tetra Pak (used for Mott's, Hawaiian Punch and Gatorade) with falsely claiming in newspaper ads that their containers can be recycled "as easily as this page." Both companies have since stopped running the ads.

Your best bet is to look for brands that specify the percentage of recycled material, such as Fort Howard and Marcal paper products, which contain 100 percent recycled pulp. If you have trouble finding 100 percent recycled products, try a mail-order house such as Real Goods (800-762-7325) or Seventh Generation (800-444-7336).

4. Dismiss those labels that say "degradable" or "biodegradable." At best the claims are meaningless, since nearly everything degrades eventually. Sometimes, however, as the crumbling Great Wall of China attests, you need to wait thousands of years. If your garbage, like that of most people, ends up buried in landfills—where no air, water or light penetrates—nothing will degrade for decades. And materials that degrade quickly aren't necessarily safe for the environment. For example, some substances, like detergents, emit toxic chemicals as they break down, warns Debra Lynn Dadd, author of *Nontoxic, Natural and Earthwise* (Jeremy P. Tarcher).

Many state and local environmental officials consider claims of degradability to be inaccurate. Last year, seven state attorneys general sued Mobil for false advertising because of claims on its Hefty garbage-bag boxes that the bags were degradable. Mobil has since agreed to settle the suits with no admission of any wrongdoing.

Environmental experts say the best thing you can do to limit waste is to avoid creating unnecessary garbage in the first place. Use cloth shopping bags instead of throwaway paper or plastic, for example. Steer clear of disposable products such as plastic razors or Styrofoam cups and opt for long-lasting, reusable alternatives.

5. Ignore "ozone friendly" spray-can labels and consider pump bottles. Many brands of aerosol shaving cream and deodorant now boast that they contain no CFCs, or chlorofluorocarbons, which harm the ozone layer—the blanket protecting the earth from dangerous ultraviolet radiation from the sun. Big deal. The U.S. government banned CFCs from aerosols in 1978. Aerosols still contain hydrocarbons, such as propane, which contribute to smog. Some even contain HCFC, a CFC-like substance that many environmentalists maintain also threatens to deplete the ozone layer.

To avoid these chemicals, use pump bottles or, to get a continuous spritz, use the new nonaerosol sprays, such as the $7.99 Biomat bottle from Biomatik (800-950-6478). Unlike the aerosol cans commonly used for hair sprays and deodorants, the new spray bottles don't require chemical propellants. They can be filled and refilled with liquids such as window cleaner or water-based paint. To create a chemical-free spray, you add compressed air by pumping up a device attached to the underside of the bottle.

6. Rather than relying on environmental "life cycle" analyses, search instead for energy-conserving products whose manufacturers have proof that using the items will cut your utility bills. Makers of disposable diapers and a trade group for cloth diaper services commissioned rival studies that attempted to measure the environmental impact of their products over their lifetimes, from manufacture to disposal. Not surprisingly, the studies drew opposite conclusions. The non-profit group Green Seal, headed by Earth Day organizer Denis Hayes, will soon begin awarding life-cycle endorsements of paper products. Hayes says that over the next few years consumers will find the Green Seal on everything from soaps to water-saving devices.

The entire-life-cycle concept may not be valid, though. Many scientists consider such analyses to be unreliable.

Parts of this listing are from the *Conservation Directory* of the National Wildlife Federation. Reprinted by permission.

African Wildlife Foundation
1717 Massachusetts Ave. NW
Washington, DC 20036
(202) 265-8394

Helps conserve natural resources through programs that focus on education, wildlife parks and special protection efforts for endangered species including mountain gorillas, rhinos and elephants.

American Farmland Trust
1920 N St. NW, Suite 400
Washington, DC 20036
(202) 659-5170

Protects America's farmlands by stopping the loss of productive farmland and promoting farming practices that lead to a healthy environment.

American Rivers
801 Pennsylvania Ave. SE, Suite 303
Washington, DC 20003
(202) 547-6900

Works to protect wild, natural and free-flowing rivers and their landscapes and acts as an information resource for activists. Works on legislation and assists local and state organizations in conservation projects.

Americans for the Environment
1400 16th St. NW
Washington, DC 20036
(202) 797-6665

Teaches citizens today to protect the environmental future for our children. Empowers Americans to safeguard land, water and wildlife by voting for conservation candidates.

Center for Marine Conservation
1725 DeSales St. NW, Suite 500
Washington, DC 20036
(202) 429-5609

Protects dolphins, whales and other endangered marine species; conducts programs to keep our nation's oceans and beaches clean; advocates for marine sanctuaries to preserve critical habitats.

Chesapeake Bay Foundation
162 Prince George St.
Annapolis, MD 21401
(410) 268-8816 (Annapolis)
(410) 269-0481 (Baltimore)
(301) 261-2350 (Washington)

Goal is to promote and contribute to the orderly management of the Chesapeake Bay with a special emphasis on maintaining a level of water quality that will support the bay's diverse aquatic species.

Clean Water Action Project
1320 18th St. NW
Washington, DC 20036
(202) 457-1286

Works for clean, safe, affordable water, control of toxic chemicals and the protection of natural resources. Provides technical and organizing assistance to groups involved in fighting incinerators, cleaning up dumps and protecting groundwater.

The Conservation Fund
1800 N. Kent St., Suite 1120
Arlington, VA 22209
(703) 525-6300

Buys land outright—60,000 acres since 1986—for wildlife refuges, parks, Civil War battlefields; also riversides, wetlands, greenways.

Conservation International
1015 18th St. NW
Washington, DC 20036
(202) 429-5660

Rescues rain forests and endangered species through wildlands management and scientific field projects. Helps indigenous people and local communities improve their lives without destroying their forests.

The Cousteau Society
930 W. 21st St.
Norfolk, VA 23517
(804) 627-1144

An international, membership-supported, nonprofit organization dedicated to the protection and improvement of the quality of life, with an emphasis on the marine environment.

Defenders of Wildlife
1244 19th St. NW
Washington, DC 20036
(202) 659-9510

A national nonprofit organization whose goal is to preserve, enhance and protect wildlife and preserve the integrity of natural ecosystems.

Ducks Unlimited
One Waterfowl Way
Long Grove, IL 60047
(708) 438-4300

Raises money for developing, preserving, restoring and maintaining waterfowl habitat in North America and educates the public concerning wetlands and waterfowl management.

Earth Island Institute
300 Broadway, Suite 28
San Francisco, CA 94133
(415) 788-3666

Initiates and supports international projects for the protection and restoration of the environment. Current focus: stopping the killing of dolphins, rain forest protection, sea turtle restoration and fostering environmental protection and restoration in Central America.

Environmental Action/Environmental Action Foundation
6930 Carroll Ave., Suite 600
Takoma Park, MD 20912
(301) 891-1100

Works for enactment of the strongest possible environmental laws and publishes the bimonthly *Environmental Action* magazine. Its affiliated political action committee, EnAct/PAC, endorses pro-environment candidates and spotlights the worst members of Congress through the "Dirty Dozen" campaign.

Environmental Defense Fund, Inc.
257 Park Ave. South
New York, NY 10010
(212) 505-2100

Work spans global issues such as ocean pollution, rain forest destruction and the greenhouse effect. Since its founding in 1967 in the effort to save the osprey and other wildlife from DDT, EDF has used teams of scientists, economists and attorneys to develop economically viable solutions to environmental problems.

Environmental and Energy Study Institute
122 C St. NW, Suite 700
Washington, DC 20001
(202) 628-1400

Independent nonpartisan organization mobilizes Congressional leadership for healthy environment and healthy economy. Seeks incentives for energy efficiency, water conservation, clean transportation, pollution prevention, international cooperation.

Friends of the Earth
218 D St. SE
Washington, DC 20003
(202) 544-2600

Dedicated to the conservation, protection and rational use of the earth. Focus is on saving the ozone layer, ending tropical deforestation, fighting global warming, tackling the waste crisis, protecting the oceans, encouraging corporate responsibility for the environment, ending nuclear weapons production and redirecting tax dollars to the environment. Recently merged with the Oceanic Society and the Environmental Policy Institute.

Global Tomorrow Coalition
1325 G St. NW, Suite 1010
Washington, DC 20005-3104
(202) 628-4016

Unites over 100 nongovernmental organizations, educational institutions and corporations in a national network. Goal: to broaden public understanding of the significance of long-term global trends in population resources, environment and development and to help promote informed and responsible public choices.

Greenpeace, USA, Inc.
1436 U St. NW
Washington, DC 20009
(202) 462-1177

Dedicated to protecting the natural environment. Works to shield the environment from nuclear and toxic pollution, halt the slaughter of whales and seals, stop nuclear weapons testing and curtail the arms race at sea. Also campaigns against the mining and reprocessing of nuclear fuel, the exploitation of Antarctica and the destruction of marine resources through indiscriminate fishing.

Inform
381 Park Ave. South
New York, NY 10016
(212) 689-4040

Researches practical strategies to improve environmentally damaging corporate and municipal practices, shaping chemical hazards and solid waste prevention programs for government, business and grassroots leaders.

International Institute for Energy Conservation
420 C St. NE
Washington, DC 20002
(202) 546-3388

Counteracts air and water pollution and the threat of global warming by working with developing countries to establish sustainable growth with efficient uses of energy.

Izaak Walton League of America
1401 Wilson Blvd., Level B
Arlington, VA 22209-2318
(703) 528-1818

Since 1922, works toward the protection of America's land, water and air resources. Current focus: acid rain, clean water, stream protection, Chesapeake Bay cleanup, outdoor ethics, public land management, soil erosion, clean air and waterfowl/wildlife protection.

Land Trust Alliance
900 17th St. NW, Suite 410
Washington, DC 20006
(202) 785-1410

Helps independent local grassroots groups preserve the open spaces that are special to communities across America—wetlands, farmland, parks, forests, wildlife habitat—for future generations.

National Audubon Society
950 3rd Ave.
New York, NY 10022
(212) 832-3200

Conserves plants and animals and their habitats; promotes rational strategies for energy development and use, stressing conservation and renewable resources; protects life from pollution, radiation and toxic substances; furthers the wise use of land and water; seeks solutions for global problems involving the interaction of population, resources and the environment.

National Coalition against the Misuse of Pesticides
701 E St. SE, Suite 200
Washington, DC 20003
(202) 543-5450

Serves as a national network committed to pesticide safety and the adoption of alternative pest-management strategies that reduce or eliminate a dependency on toxic chemicals.

National Parks and Conservation Association
1776 Massachusetts Ave. NW, Suite 200
Washington, DC 20036
(800) NAT-PARK

Preserves national parks from Grand Canyon to Gettysburg; protects endangered wildlife, majestic scenery, cultural sites; promotes new parks; defends against overuse, development and other threats.

National Wildlife Federation
1400 16th St. NW
Washington, DC 20036-2266
(202) 797-6800

Promotes the wise use of natural resources and protection of the global environment. Distributes periodicals and educational materials, sponsors outdoor

education programs in conservation and litigates environmental disputes in an effort to conserve natural resources and wildlife. Current focus: forests, energy, toxic pollution, environmental quality, fisheries and wildlife, wetlands, water resources, public lands.

Natural Resources Defense Council
40 W. 20th St.
New York, NY 10011
(212) 727-2700

Combines legal action, scientific research and citizen education in environmental protection program. Major accomplishments have been in the areas of energy policy and nuclear safety, air and water pollution, urban environmental issues, toxic substances and natural resources conservation.

The Nature Conservancy
1815 N. Lynn St.
Arlington, VA 22209
(703) 841-5300

Committed to preserving biological diversity by protecting natural lands and the life they harbor. Works through 50 state offices to identify ecologically significant natural areas. Manages a system of over 1,600 nature sanctuaries nationwide. Works with Central and South American conservation organizations to identify and protect natural lands and tropical rain forests.

Pesticide Action Network
965 Mission St., #514
San Francisco, CA 94103
(415) 541-9140

Works to eliminate poisonous pesticides worldwide. Links local consumer, worker, environmental and agriculture groups internationally to advance safe, sustainable pest control.

Rails-to-Trails Conservancy
1400 16th St. NW, Suite 300
Washington, DC 20036
(202) 797-5400

Works to convert thousands of miles of abandoned railway corridors into public trails for walking, bicycling, horseback riding, cross-country skiing, wildlife habitat and nature appreciation.

Rainforest Action Network
450 Sansome St., Suite 700
San Francisco, CA 94111
(415) 398-4404

Works to save the world's rain forests. Works internationally in cooperation with other environmental and human rights organizations on campaigns to protect rain forests. Provides financial support for groups in tropical countries to preserve forest lands. Methods include direct action, grassroots organizing and media outreach.

Rainforest Alliance
270 Lafayette St., Suite 512
New York, NY 10012
(212) 941-1900

Conserves endangered tropical forests through community-based field projects, public education and research. Develops opportunities for people to utilize tropical forests without destroying them.

Renew America
1400 16th St. NW, Suite 710
Washington, DC 20036
(202) 232-2252

A national clearinghouse for environmental solutions, this organization fosters the rapid, efficient expansion of successful environmental programs and encourages cooperation and consensus among environmental interests. Oversees two national programs: "Searching for Success" collects and evaluates data on working environmental programs. The "State of the States" program compares and ranks state environmental efforts.

Rocky Mountain Institute
1739 Snowmass Creek Rd.
Snowmass, CO 81654-9199
(303) 927-3851

Mission is to foster the efficient and sustainable use of resources as a path to global security. It focuses its efforts in five program areas—energy, water, agriculture, economic renewal and security.

Safe Energy Communication Council
1717 Massachussets Ave. NW, Suite LL215
Washington, DC 20036
(202) 483-8491

Fights for environmentally safe, affordable energy. Promotes renewable energy such as solar, wind-power and energy efficiency. Informs communities about nuclear power and waste dangers.

Scenic America
21 DuPont Circle NW
Washington, DC 20036
(202) 833-4300

Preserves and enhances America's scenic resources and combats visual pollution. Provides technical assistance and research empowering citizens nationwide to protect community character and scenic landscapes.

Sierra Club
730 Polk St.
San Francisco, CA 94109
(415) 776-2211

Promotes conservation of the natural environment by influencing public policy, practices and promotes the responsible use of the earth's ecosystems and resources and educates and enlists humanity to protect and restore the quality of the natural and human environment.

Tarlton Foundation
50 Francisco St., Suite 103
San Francisco, CA 94133
(415) 433-3163

Aids in the preservation of the earth's waters through conservation-focused education and research programs. Programs include Project OCEAN, offering elementary and middle school teachers in-service workshops and summer institutes, and free access to a marine science library. Also sponsors the Whalebus, a hands-on outreach program; Sea Camp, a summer marine science program for children ages 6 to 14; and the Adopt-a-Whale and Adopt-a-Beach programs.

Trout Unlimited
800 Follin Ln. SE, Suite 250
Vienna, VA 22180
(703) 281-1100

Dedicated to the protection of clean waters and the enhancement of trout, salmon and steelhead habitat. The national office works with Congress and federal agencies for the protection and wise management of America's fishing waters; sponsors seminars; funds research projects; and administers the nationwide network of chapters.

The Trust for Public Land
116 New Montgomery St., 4th Floor
San Francisco, CA 94105
(415) 495-4014

Acquires and protects open spaces for people to use and enjoy as urban parks, neighborhood gardens, trailways, greenways and recreational areas.

Union of Concerned Scientists
26 Church St.
Cambridge, MA 02238
(617) 547-5552

Advocates energy strategies that minimize risks to public health and safety, provide for efficient and cost-effective use of energy resources and minimize damage to the global environment.

U.S. Public Interest Research Group
215 Pennsylvania Ave. SE
Washington, DC 20003
(202) 546-9707

Represents the public's interest in the areas of consumer and environmental protection, energy policy and governmental and corporate reform. Current focus: clean water, solid waste, energy efficiency, antinuclear activities, global warming, ozone depletion and reducing the use of toxics.

The Wilderness Society
900 17th St. NW
Washington, DC 20006
(202) 833-2300

Since 1935, devoted to preserving wilderness and wildlife, protecting America's forests, parks, rivers and shorelands. Welcomes membership inquiries, contributions, publication requests and bequests. Also houses 75 original Ansel Adams photographs in a custom-built gallery.

World Resources Institute
1709 New York Ave. NW, 7th floor
Washington, DC 20006
(202) 638-6300

Helps address the fundamental question of meeting human needs and nurturing economic growth while preserving natural resources and environmental integrity. Current focus: tropical forests, biological diversity, sustainable agriculture, energy, climate change, atmospheric pollution, benign environmental technologies, international institutions and economic incentives for sustainable development.

World Wildlife Fund/The Conservation Foundation
1250 24th St. NW
Washington, DC 20037
(202) 293-4800

Works to preserve biological diversity through wise resource management, habitat protection and sound environmental practices. Current focus: creating sustainable development programs in Latin America, Asia and Africa; assisting nongovernmental organizations in developing countries; planning and implementing debt-for-nature swaps in Latin America, Asia and Africa; and monitoring international wildlife trade.

Zero Population Growth
1400 16th St. NW, Suite 320
Washington, DC 20036
(202) 332-2200

Works to achieve a sustainable balance between the earth's population, its environment and its resources.

INDEX